Ledyard Bill

A Winter in Florida

Observations on the Soil, Climate, and Products of our Semi-tropical State

Ledyard Bill

A Winter in Florida
Observations on the Soil, Climate, and Products of our Semi-tropical State

ISBN/EAN: 9783744757409

Printed in Europe, USA, Canada, Australia, Japan

Cover: Foto ©berggeist007 / pixelio.de

More available books at **www.hansebooks.com**

A

WINTER

IN

FLORIDA;

OR,

OBSERVATIONS ON THE SOIL, CLIMATE, AND PRODUCTS OF OUR
SEMI-TROPICAL STATE; WITH SKETCHES OF THE PRIN-
CIPAL TOWNS AND CITIES IN EASTERN FLORIDA.

TO WHICH IS ADDED

A BRIEF HISTORICAL SUMMARY;

TOGETHER WITH

HINTS TO THE TOURIST, INVALID, AND SPORTSMAN.

By LEDYARD BILL.

ILLUSTRATED.

SECOND EDITION.

NEW YORK:
PUBLISHED BY WOOD & HOLBROOK,
13 & 15 LAIGHT STREET.
1869.

BOSTON:
RAND, AVERY, & FRYE, STEREOTYPERS AND PRINTERS,
NO. 3, CORNHILL.

THIS VOLUME

IS CORDIALLY INSCRIBED

TO

C. L. R.,

BY THE AUTHOR.

PREFACE.

THE tour of which this volume is the record was an unforeseen one, the sudden illness of a friend having been the occasion of our visit to Florida.

Diligent inquiry brought to light no work on the State, whereby we might be guided. Beyond several historical summaries, given the public a quarter of a century since, there was absolutely nothing worthy of mention; and even these were valueless except to the historical student, anxious only to know its colonial history and status up to the period of its acquisition

by the United States. To in some measure
supply this deficiency has been the object
of our labors.

Florida is the oldest settled portion of
the Union, notwithstanding which it is the
most of a wilderness. Especially is this true
of the eastern half, which was the portion
visited by us, and to which we have in the
main confined ourselves.

The history of the State is renowned,
from its first settlement to its annexation,
for its many battles and conquests, in which
the victors of to-day were among the van-
quished on the morrow; and thus its masters alternated for centuries, leaving little
else behind them than vestiges of their
occupation. But, since our Government has
held possession, much has been accomplished
in material prosperity and in increase of
population : still, three-fourths of the land

is apparently as wild and unoccupied as ever.

Within, however, the past few years, Florida has attracted considerable attention as a winter resort for invalids and pleasure-seekers. It is, practically, the only strip of tropical land within our boundaries, and the only State where the invalid can find an equable and mild temperature through the greater portion of the year. Visitors to the State are already numbered by thousands, and each year since the war has witnessed a rapid increase.

The St. John's, whose source is in the everglades of central and southern Florida, running due north for two hundred miles, then abruptly turning eastward to the ocean, is in many respects the most remarkable river in North America. To the sportsman it presents opportunities such as no other offers, while the pleasure-seeker is amply repaid in viewing its unusual beauties.

The capabilities of the State in an agricultural point of view are unbounded. The growth of coffee, cotton, and cane, as also indigo, and not unlikely tea, together with the production of wine, is likely to form an important chapter in her future history; but the thing above all others in which Florida is certainly destined to excel her sister States is in early fruit-growing and marketing. The climate is favorable for most of the tropical fruits, as also for those grown in more northern latitudes. The soil, too, is admirably adapted to their rapid and early maturity. Already a large share of attention is given this department of agriculture, both from the native and immigrant population; and we hear, that, in the matter of oranges alone, over half a million trees

have been set out along the St. John's
and its tributaries within the last twelve
months.

There are many advantages to any
people who have more than a single interest in which to enlist their energies. Hitherto the South has had but one great crop,
and that was cotton. It was profitable to
the large planter ; and its tendencies were
a concentration of wealth and the elevation
of the few : whereas a diversified industry
among a people not only makes them more
independent, but greater enlightenment
follows. There is no country on the globe
where this is so remarkably exemplified as
in "sterile" New England. The whole
Southern country to-day needs, more than
all things else, a broader culture both in the
field and in the school. This accomplished,
together with an entire revolution of her

old-time intolerance and exclusiveness, and
her road to independence and importance
will be found both broad and easy.

June 30, 1869.

TABLE OF CONTENTS.

11

CHAPTER VI.

FROM JACKSONVILLE TO GREEN-COVE SPRINGS.

CHAPTER VII.

CENTRAL FLORIDA.

CHAPTER VIII.

THE UPPER ST. JOHN'S.

CHAPTER IX.

CELEBRATED SPRINGS.

CHAPTER X.

ALLIGATOR-SHOOTING ON THE UPPER ST. JOHN'S.

CHAPTER XI.

ST. AUGUSTINE.

CHAPTER XII.

THE CLIMATE.

CHAPTER XIII.

THE SOIL.

CHAPTER XIV.

FRUITS.

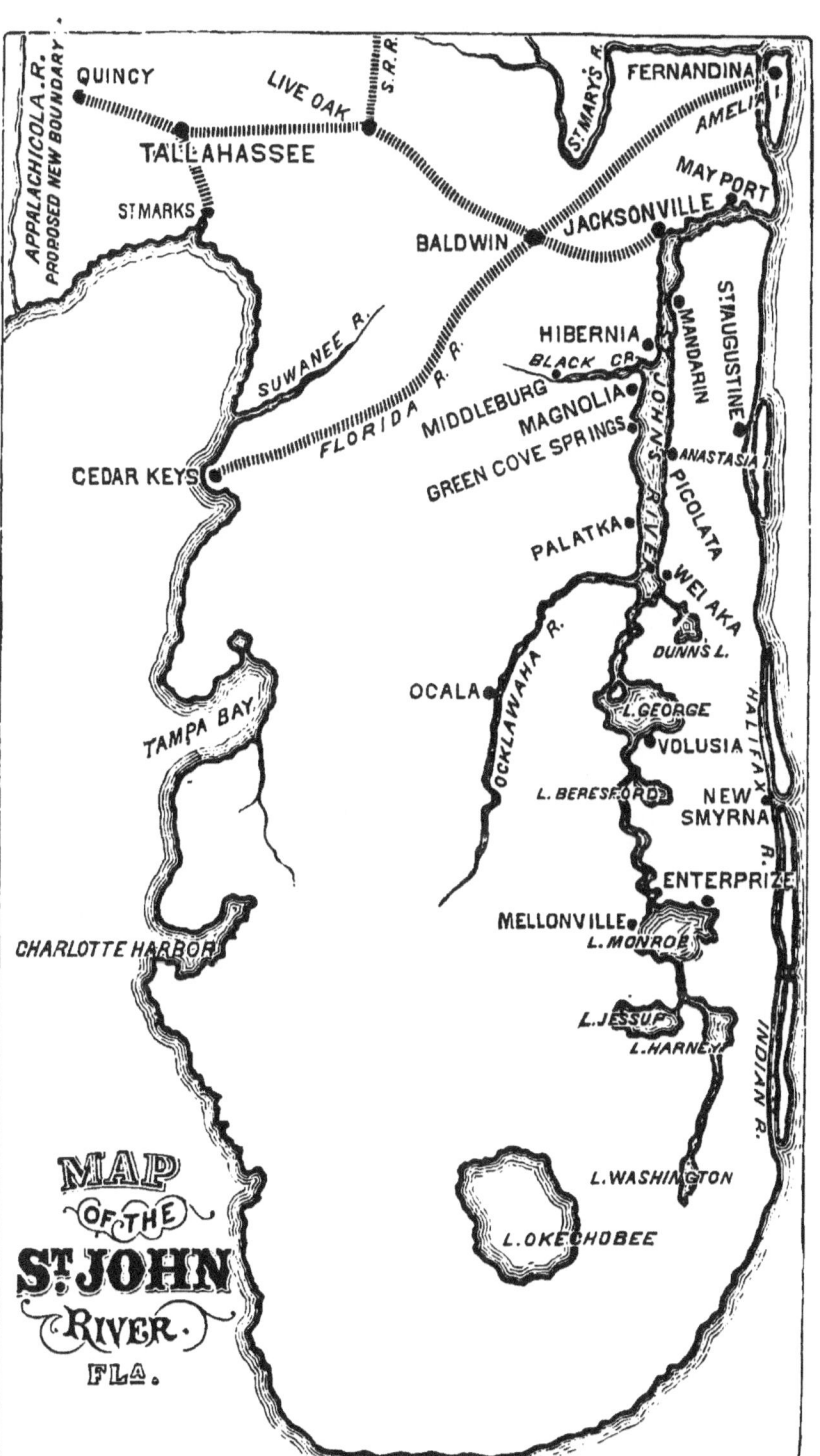

MAP OF THE St. JOHN RIVER FLA.

A WINTER IN FLORIDA.

INTRODUCTORY CHAPTER.

WHAT TO EXPECT, AND HOW TO GO.

The Mania for Travel. — Contrast in the Modes of Living. — The
South behind the Age. — Absence of Home Comforts. — Florida
a Wilderness. — Luxuries from the North. — How to go to Florida.
— The Best Routes. — Benefits of a Sea-Voyage.

THE visitor to Florida is likely to be disappointed
in many respects. The exchange of the comforts
of a home for a temporary residence in any remote
portion of our country is always effected at a sac-
rifice ; and those who undertake journeys, whether
for pleasure or health, have to learn the lesson of
circumscribing their wants, and, not unlikely,
many of their necessities. How many of our
fashionable people feel it imperative to arrange
some excursion by rail and river in midsummer to

17

the over-packed hotels, where they are bundled
into the fifth story, and their families crammed into
one or two confined and heated rooms, in exchange
for their cool and ample apartments at home. No
complaint is made in these cases, since fashion
dictates ; but this weak and amiable dame has as
yet scarce heard of Florida. If this periodical
hegira was for the purpose of escaping any ill, then
it would be tolerable ; but they fly to a thousand
they have little knowledge of. In this perhaps,
after all, compensation may be found, since the
monotony of life is broken ; and, by being divorced
from abundant and self-satisfying supplies, people
are better prepared to return to their enjoyment
with an appreciating relish.

If, in lieu of this watering-place mania, it were
the annual custom of the people of the North to
take their flight with the birds, and leave behind
them the chilling winds and frosts of winter, for
sunnier skies, where the wild-flowers and songs of
birds are perennial, then the object could not be
doubtful, and the advantages so obscure.

There is in humanity an innate desire to follow
the retreating sun, and live beneath shadowing
trees, and slake the thirst from the cool waters of

unfrozen streams; but the laws of *labor* and of our civilization are such as to make it impossible but for the few to avail themselves of these privileges.

The deprivations attending a journey into any sparsely-settled State are seldom considered. The prospective pleasures are alluring, while the realities are unknown. In nearly all southern countries, whether in the New or Old World, there is a great dearth of what northern people know as " home comforts." Instead of the snug and neat dwelling, a plain and often rude structure is seen ; but this is, after all, to the advantage of health, since it gives ample ventilation, and in all warm countries this is a very great desideratum. It is doubtful if the inhabitants of a tropical country would survive a single generation in our air-tight dwellings. But the contrast to the visitor is naturally great, and is not viewed except with some slight alarm. The universal indolence of southern populations is proverbial, and is attributed to the effects of the enervating climate in which they reside, as also to the ease which attends the cultivation and growth of such products as are necessary to their inexpensive modes of living.

The South is full a hundred years behind the

North in many things; and this is more true when speaking of Florida than of the other Southern States. It is a sort of a wild pasture-ground, parceled out into great estates, several thousand acres each, under the old Spanish grants. Not a tenth of the land is yet cleared, while the remaining portion is but a tangle of swamp and pine-forest, interspersed with lakes and rivers. Great herds of cattle roam unrestrained over the State, being gathered in by the owner perhaps once a year, for the purpose of marking the increase, and again turned loose. During the late war, thousands of these cattle were slaughtered, both by the Union and Rebel forces, and often were the only source of supplies. The loss of property to the State in this way alone must have amounted to several millions of dollars.

Notwithstanding the quantities of nearly wild cattle, there is not, so far as our observation went, a single first-class milch cow in all Eastern Florida. Why the inhabitants should pay so little attention to what would contribute so much to their comfort, is a question not answered by them. Perhaps, now that the State is being invaded by a class of people who will clamor for these ordinary

comforts, — luxuries, we should say, for good milk
and butter is more or less a luxury in Florida, —
the inhabitants may wake up to their necessities.
Poor milk, and a great scarcity of that, is the
rule wherever we traveled.

There are no good pastures at present for cattle :
the wild grasses and shrubs on which this present
neat stock subsist would not do for dairy stock ; and
whether or not the soil and climate are favorable
to the growth of northern grasses, we cannot say,
though we see no reason why clover should not
grow luxuriantly. There is one way, at least, as
practicable there as anywhere, and that is stall-
feeding; but that would be an extravagance and
a labor unheard of. All the butter used is brought
from the North, chiefly from the New-York
market. Beef is also imported, but generally
from some of the Southern cities with which steam
communication is had. Florida beef we met with
occasionally, which was very good, being both
sweet and tender; though this does not tally with
the experience of others who have traveled the
State.

Venison and various kinds of wild game abound,
together with the finest of fish, that may be had

from every river and lake. These, with what of
luxuries are imported from the North in the way
of butter, canned fruits, and ice, make up a variety
of fare that should satisfy every appetite, however
pampered it may be. It was our universal
experience, whether on the river steamers, or at
the hotel or boarding-house, that the table was
unexceptionable : we mean in the traveler's sense
and view of these things.

The cost of living in Florida, for a family,
would be thought much cheaper than in the more
northern States. This is doubtless true, so far as
the native population are concerned ; but the
families from the North who continue their
previous style would likely find it quite as expensive, since many of the articles are procured
from the same markets as when at home. By and
by this will be greatly changed, and to the advantage of the people of Florida, as well as to those
who temporarily sojourn there ; as where now little
beside the wilderness is seen, must in a few years,
in all probability, be found the prosperous homes
of a new population.

It is in fact only since the war, that Eastern
Florida has been truly opened to the North. The

former system of espionage upon all Northerners, and their certain ostracism by the chivalry, was even more effectual in its results than any Chinese Wall could possibly be.

A brief mention of the various routes by which the invalid or tourist may reach the State seems desirable. The *cheapest* method is to embark in some sailing-vessel leaving New York, Boston, or Philadelphia, direct for Jacksonville. Those who go to make the State their home would at least do well to ship all their household articles, in this way saving very much in transportation, and losing but very little in time, since the passage does not occupy over ten, and frequently is made in five days. Once at Jacksonville, direct and rapid communication is had with all portions of the State.

Those desiring to avoid a sea-voyage can go to Jacksonville by railroad the entire distance, by way of Savannah, the cars and sleeping-coaches of the South being in most respects equal to those in use in any portion of the country. Still another route, combining both land and water, is to proceed to Charleston by rail, where tri-weekly steamers run direct to the St. John's River, and up it to Palatka. There are, indeed, two lines of steamers from Charleston to Florida, one known as the "inside" and the other as the "outside" line. The former follows the bays and inland waters along the coast, touching at the small towns and landings along the seaboard, and is nearly sheltered from the ocean. The latter line goes directly out to sea from Charleston, touching at Savannah and Fernandina only, but making every landing on the St. John's as far as Palatka. A day in Charleston, viewing the "hot-bed," and a sail down the bay to old Fort Sumter, where Anderson and his brave band stood so nobly, is not lost; and a night at the commodious and well-kept Charleston Hotel gives rest and refreshment for the balance of the journey.

But the easiest and most expeditious way is, however, by some one of the various steamship lines from New York, Philadelphia, or Baltimore. They average at least one boat per day. We regard these lines as in many respects the best means of access to Florida, whether for the invalid or excursionist. The time occupied is less, and the expense is also less; and they run first-class ocean steamers, with every accommodation required. The passage to Savannah is usually made inside of three days, and direct connection is made with some one of the steamship lines to Florida; or, if parties choose, they can, by taking the cars at Savannah, reach Jacksonville in twelve hours, making the entire trip from New York in from three to four days.

The sea-voyage, dreaded by some, is of short duration, and in most cases a decided benefit to all, and especially the invalid. It is frequently recommended by physicians to those whose health makes it necessary to visit a warmer climate. There need be no fear of taking cold, for few people ever take cold from any exposure to the open sea-breeze: it is, on the contrary, invigorating and healthful.

We have elsewhere stated with greater particularity the time visitors to Florida should start. October is as early as Northern people need to go, and they may remain as late as May. Of course we now refer to special classes, who go for pleasure, and those seeking a restoration of their health. Northern people may and do remain the year round, with perhaps as much immunity from disease as when at home.

CHAPTER II.

DISCOVERY OF FLORIDA.

FLORIDA was first discovered by that bold and
early navigator, Sebastian Cabot, who, with his
father, John Cabot, in 1497, coasted the whole of
the eastern shore of the peninsula ; and, so far as
history authenticates, theirs was the first landing
made by any Europeans on this continent. Every
child is taught, however, that to Christopher Columbus belongs the honor of the discovery of the
New World, in 1492, five years prior to the landing of Cabot in Florida ; but his discoveries at that
time embraced only a portion of the Bahama
Islands. He did not touch upon the main until

the year 1498, when he landed at the mouth of the Orinoco in South America; and, even then, he did not know that his feet pressed the soil of a new hemisphere, but, on the other hand, believed that he stood upon the eastern shores of Asia: and in this faith he died, not even dreaming of the real grandeur of the discovery. Both Columbus and Cabot were, however, robbed to a great extent of their due share of honor, by a subsequent navigator, who was more fortunate in giving a description of his voyages, which was published in the Old World about the year 1500.

Little further was known of Florida by any of the Europeans until the spring of 1512, when Juan Ponce de Leon, an aged Spaniard, sailed from Porto Rico in search of a fountain reputed to exist among the Bahama Islands, which would restore the aged to immortal youth, if they but drank of its sparkling waters. Of this romantic old man of the sea, all have read in history. That he had a poetic frenzy, bordering on extravagance, the reader will scarcely question. If he had lived in the present age, he would, doubtless, have been sent away, by either his heirs-at-law or his darling children, to contemplate Nature from behind some

Gothic window, carefully protected against intrusion from either burglars or Bohemians.

His career is indeed fascinating in its contemplation; and to think of him setting sail from old Spain, in that early age of navigation, across the far-reaching and stormy waters of the Atlantic, in his rude vessel, surrounded by a few devoted, and, we must presume, unlettered sailors, challenges our admiration, and we follow his subsequent career with unabating interest.

The object of his search was like the mirage in the desert to the worn and weary traveller, who is beguiled to pursue its ever-retreating cool fountains and inviting groves, till at last he sinks exhausted, yet believing that the object of his search lies nearly within his grasp.

De Leon, after a fruitless visit to the Bahama Islands, steered north-west for the coast of the continent, and landed at what is now St. Augustine, still pursuing the object of his industrious search. He landed in April, 1512, on Easter Sunday, which day the Spaniards call *Pascua Florida;* and, finding the country abounding in wild-flowers, he gave it the name of Florida: which is, indeed, a very beautiful name for a beautiful land. He at once arranged for an active campaign of discovery for the one absorbing object of his ambition. By land, lagoon, and river he traveled, suffering hardships and enduring every thing. Thus he journeyed and coasted southward the whole extent of the peninsula, even to the Tortugas Islands in the Gulf, whence he returned to Spain, disappointed but not entirely disheartened; for, in a few years, we find him commissioned by the home authorities as governor of Florida, conditioned on his establishing a colony in that country. This enterprise he readily undertook; and, gathering a company of emigrants, he set sail and effected a landing, but was met by the aborigines with frowns and suspicions, which speedily grew into open and determined hostility, resulting in the killing and driving of the colonists from the shores to their ships. From that day, now over three hundred and fifty years, one continuous battle with the red men of the forest has been waged, with brief respites only intervening, till they now stand at bay two thousand miles in the interior of the continent, cut off, surrounded, and hunted like wild beasts, they and their history soon to be swept away by the great wave of civi-

lization which has fully set in along the iron road
across the continent to the ocean beyond.

In this first conflict with the red men in Florida,
De Leon was mortally wounded, and soon after
died in Cuba, whither the expedition had set immediate sail. Thus did this romantic voyager's
life abruptly and tragically end. There is scarcely
in history a character so unique as his. That he
was a man of influence among the rulers of Spain,
and was possessed of remarkable parts, can not be
questioned; nor the fact of his many amiable
weaknesses, which seemed to exert a controlling
influence over him.

A dozen years later, and the Spanish Government had appointed one Narvaez as ruler over
this wild-flowery land of De Leon's. He sailed
from Cuba with a force of three hundred men,
well armed and equipped, to take forcible possession, and hold the vast territory which was now
regarded as so valuable to the Spanish crown.
His expedition attempted little of colonization, but
gave their time chiefly to the exploration of the
interior country. They were beset on all sides in
their marches, and at last so hunted and harassed
by the Indians, and cut down by sickness, that

only four who survived were able to reach a friendly settlement, hundreds of miles away in Mexico;
and this only after several years of wanderings.
De Leon's dreams still haunted the minds of adventurers, changing only in that gold and precious
gems took the place of his fountain which was to
impart the gift of perpetual youth.

In this age of romantic and chivalrous adven-
ture, of new-found worlds and empires, it is not so
wonderful, perhaps, that eager knights embraced
almost any opinion concerning the mysterious
recesses of lands which now lay open to their view.
The discovery and conquest of Peru preceded the
discovery of Florida but a brief period; and the
wealth which fell to its captors but added to the en-
thusiasm that already stirred the breasts of the
people of Spain.

Ferdinand de Soto, a principal actor in this
grand drama, who had returned to Spain loaded
with fame and wealth, at once resolved on an expedition to Florida, and sought permission to con-
quer and govern the territory, which was granted.
Rallying around him great numbers of adventurers
and admirers, he selected from among them about
a thousand men suited to his purposes, and on

the sixth day of April, 1538, set sail in a fleet of
ten vessels. Arriving at Cuba, where he was
joined by Porcalho, a distinguished and wealthy
soldier of fortune, he was obliged to repair and
re-form. This occupied nearly a year; and it was
not until the following spring of 1539 that he
landed on the shores of Tampa Bay, in Florida.

It was the sincere wish of De Soto to conciliate
the inhabitants, and open the way to a peaceful
exploration of the country. In fact, the measures
he adopted were admirably adapted to that end;
but the natives well remembered the cruelty of
Narvaez, and they regarded the new comers as but
another band of robbers.

De Soto sent out couriers to trace, if possible,
the fragments of the preceding expedition; but
this proved fruitless, since they were invariably
detained, and usually put to death. Finding all
measures of conciliation unavailing, he resolved
on conquering the country by force of arms.
To this, Porcalho dissented, and, unable to effect
an agreement, sailed away; and immediately De
Soto set out on his perilous and disastrous march.
Had the native population possessed an organized
body of men, and a seat of government, and pros-

ccuted hostilities any thing like ordinary mortals, they would have been overpowered and speedily reduced, and brought under the sway of the disciplined and hardy veterans of Spain ; but De Soto struggled against a multitude of fierce, petty tribes, who, although offering no point at which an effective blow could be struck, never left him master of more than the spot on which, for the moment, his army stood. Wherever he advanced, a wild shower of poisoned arrows greeted him, or else the foe fled before him, leaving but a barren waste behind.

The natives at length prepared a plot for the extermination of the invaders, and, professing friendship, advanced suddenly upon them, taking them wholly by surprise. The battle wavered but for a moment ; when the soldiers, having rallied with their arms, poured in a destructive fire, which killed great numbers of the enemy, and put to flight the remainder, except a few who were surrounded. These, on seeing their position, plunged into a lagoon, and there floated, exposing nothing but their faces above the water. In this position they were kept till the following day by the determined Spaniards, when they were

3

glad to surrender. They constituted, however, but a fragment of the dusky warriors who everywhere haunted the unbroken forest.

Florida was at this time without definite boundaries. At the time of De Leon's expedition, or soon after, the Spanish Crown claimed authority from Cape Sable up as far as Labrador; but their expeditions seldom if ever extended farther up the coast than Cape Hatteras, and their pretensions to territory above the Carolinas was abandoned; but up to that point they laid strenuous claims.

Remembering his great good fortune in Peru, and still dazzled by the wealth and splendors of that conquest, De Soto's thoughts still dwelt on some sunny clime in Florida, where temples and princely palaces glittered with burnished silver and gold, where sparkling gems gave back with added brilliancy the slanting sunlight, and where the very air was soft and fragrant with aromatic and priceless gums and balsams.

Weary and worn, yet undaunted, this brave and gallant knight of the middle ages of chivalric and adventurous Spain set his face toward the wilderness, and resumed his perilous marches across low lands and rivers, through forests and by foes,

back into the unknown and heretofore unexplored
regions of America. Passing at first northward,
he traversed what is now Georgia, and then, turning
south and westward, marched to Mauvilla (Mo-
bile), in what is now Alabama, near which place
lived the chief of the Choctaw nation, a powerful
tribe inhabiting that region.

While entertaining him and his followers as
guests, and amidst the hilarities of the occasion,
the Indians commenced a fierce attack, compelling
the Spaniards to fall back to their camp, when,
mounting their horses, they advanced and drove
the Indians, though a cloud of arrows, it is narrated,
darkened the sky. On entering the village, they
were fired upon from all sides, till they resolved
to fire the town, which, being constructed of reeds
and branches, flamed like tinder: many were
thus destroyed, and a total loss is reported of over
two thousand Indians, while but eighteen of the
invaders were killed. This tragedy ended, and
De Soto learning of the existence of a great river
westward, he set out once more through a trackless region, and this time in search of the " Father
of Waters." Arriving in the Mississippi Valley,
he sought the capital of the Chickasaws, and spent

here the winter. The treachery of the savages again burst upon this devoted band; again were they surprised and slaughtered, losing some lives and many horses and all their baggage. They then moved forward to the banks of the great river; and for the first time, in the spring of the year 1541, the eyes of Europeans rested upon it.

Here they lingered to construct means of crossing; and then, still undaunted, the bravest soldier and explorer of any age or country moved on, penetrating the wild regions of the Missouri. But to follow and describe all his wanderings would require a volume. Three years had worn away since he landed at Tampa Bay with his brave army, which now had dwindled to one-half its original numbers. He determined to reach the coast, obtain re-enforcements, and inaugurate a new campaign. Descending the Washita and Red Rivers to their confluence with the Mississippi, he here halted to prepare for the descent of that river, when he was seized by fever, which in a few brief days terminated his intrepid career. His body was secretly and solemnly buried, at midnight, in the bosom of the great river, by a few of his chosen companions, with nothing around save the starlight night, the rolling flood, and the dark, impenetrable wilderness.

His survivors, descending the river, soon reached the coast, and returned to their native country, having lost two-thirds of their original number.

We have followed, briefly, the career of this, the third governor of Florida, and his most remarkable expedition, since it not only concerned the territory of Florida itself, but was important in other respects. Every step which De Soto advanced westward but added to its area. Thus we see Florida extended from Cape Sable to Cape Hatteras in the north, and westward more than a thousand miles, embracing all the country to and west of the Mississippi, and south of the Missouri River, including all drained by the Red and intervening rivers to the Rio Grande.

CHAPTER III.

The First Protestant Colony in America. — Enmity of the Spaniards. — Their Cruelty and Barbarism. — The Founding of St. Augustine by Menendez in 1565. — The Death of Ribault avenged by De Gourgues. — Capture of St. Augustine. — Drake's Expedition. — Interior Settlements of the Spaniards. — Trial of an Indian Chief for Treason. — Gov. Moore's and Oglethorpe's Attack on St. Augustine. — Florida ceded in 1763 to Great Britain, and re-ceded in 1783 to Spain. — Border Troubles. — Annexation of Florida to the States by Treaty in 1821.

WE have seen how three successive Spanish expeditions came and went, and yet Florida held no settlement of Europeans. The failure of the last, under its great leader, had dampened the ardor of that nation ; and at this time the Huguenots of France, persecuted at home, resolved to found a colony in the New World, and immediately put their project into execution.

Under the patronage of the French Admiral Coligny, a company of these people sailed in 1562, under the lead of John Ribault, who, landing at

Port Royal, took possession of that section in the name of King Charles (Carolus) of France, and gave the name of Carolina to the northern portion of Florida, — a name that still continues, divided between two States.   In 1564, Ribault sailed in charge of another expedition, and landed at the mouth of the St. John's, and there established a colony.   The news of this expedition, reaching Spain, aroused the Spanish people ; and the crown at once authorized one Don Pedro Menendez to proceed to Florida, being first duly appointed governor.   He departed with nearly a thousand men, direct for the coast of Florida, and, landing, established in 1565 a settlement on the present site of St. Augustine.*   His objective was the French settlement on the St. John's River.   Ribault was instinctively aware of this, it would seem, and made an effort to circumvent the Spaniards, which proved abortive.   Menendez began duly to arrange for an attack on this colony at or near the mouth of the St. John's ; and, taking some five hundred of his men, he commenced, though not without some protests, his march across the country, urging them forward in the name of

---

* Williams, the historian, gives 1564 as the date.

honor and their religion. Arriving near the French settlement, early on the morning of the fifth day, and finding the gates of the fort open, they rushed in, giving no quarter to the occupants, both citizens and soldiers, slaughtering over two hundred, and only saving some of the women and children, and not even them till their thirst for blood had been fully satiated.* A more heartless butchery is nowhere to be found in the annals of history; and no people among Europeans, even at this early period, save the Spaniards, could have committed the awful catalogue of crimes of which they were guilty.

Even in this nineteenth century their passions seem as uncontrolled as ever, of which the conflict in Cuba gives ample proof. No country has so long withstood the influence of the centuries as has Spain; but it is a consolation to contemplate the wheels of progress, and to note that the days of her priestcraft and old tyrannous system of government are passing away, while new Spain is robing herself, as we trust, in the beautiful garments which Liberty is weaving.

* A Spanish writer gives this statement of the saving of women and children, but the French never believed any were ever spared from the sword.

The infamous Menendez, when the work of assassination had ended, suspended from the neighboring trees a number of the mangled bodies, with the following words attached : " Not because they are Frenchmen, but because they are heretics and enemies of God."

When the news of this dreadful business reached France, it greatly excited the Protestant inhabitants ; but they were powerless, since their sovereign, Charles IX., did not sympathize with them. Roused by grief and rage, the Huguenots eventually found a leader in the daring warrior, Dominique De Gourgues, who had himself suffered every indignity during a period of captivity under the Spaniards. He clandestinely procured a license to engage in the slave-trade, and set sail in three ships in the summer of 1567. After sailing southward, he announced his determination and real intention to his men, which was to attack the Spanish settlement at St. Augustine, and avenge the slaughter of Ribault and his companions. At first, several ships' companies desisted ; but, being urged, they consented, and the expedition effected a landing. Before marching to the attack, he secured the co-operation of the natives, who were

only too ready to aid in the dislodgment of the
hated Spaniards. The garrison of the forts were
apprized of the rapid advance of the enemy, and
were preparing to receive them, when the assault
was commenced upon an outer fort, or block-
house, which was speedily carried, as was also the
next; but there yet remained the fort proper,
whose garrison far outnumbered the assailants.
From this fort a company of soldiers advanced to
meet the attack of the assailants, but they were
cut off and destroyed. A panic seized upon the
remainder of the garrison ; and, abandoning their
stronghold, they fled precipitately to the woods.
Numbers of them were captured. These the
French leader took to the fatal trees where the
fragmentary remains of his own countrymen still
hung, and, having reviled them for their cruelty,
caused a portion of them to be suspended, and, in
place of the former, substituted the following in-
scription : " Not because they are Spaniards, but
because they are traitors, robbers, and murderers."
Retribution had been swift and sure. Had none
but the ringleaders been executed, the world would
have justified the deed.

The conquerors, having accomplished the object

of their visit, set sail for home, and were, by their friends, joyfully received. De Gourgues was subsequently obliged to retire from his country (France) to escape the embarrassments under which this expedition had brought him; his king neither daring to imprison nor allow him to remain.

Thus was the foundation-stone of St. Augustine baptized in blood; and its subsequent history, for upwards of two and a half centuries, is one of prolonged slaughter and destruction. No other city in America has so often and so severely suffered from the ravages of fire and famine, and the general vicissitudes of fortune, as this, the first *permanent* settlement on the continent.

Menendez, perhaps unfortunately, was not among those who perished; and, after the departure of the conquerors, he resumed his rule, and continued for twelve years to preside over the destinies of the town and adjoining settlements or trading-posts. In 1578 he returned to Spain, having first nominated his successor. Spanish ambition in this portion of America had been too severely pruned by the sword for any rapid development of power or numbers: yet we find no diminution of papal fanaticism; but, on the other

hand, they now gave their attention to the conver-
sion of the aboriginal inhabitants to their super-
stitions and fanatical religious faith. Under the
patronage of the pope of Rome, the king of Spain
dispatched a large company of teachers, of the
order of Franciscans, to Florida. Thus, while enlarging the boundaries of their church, they sought
to obtain peaceful control over the tribes surrounding them, which, up to this period, had been a
fruitless effort. The savage tribes of all countries
have ever been more or less impressed with the forms
and ceremonials of the Romish Church ; and in a
few years, by their persistent and insinuating manners, coupled with lessons of civilized life, the
Spaniards acquired great influence over them, and
by 1585 their rule was acknowledged by all of the
tribes as far west as the Mississippi River. It was
during this period that those convents and mission-
ary houses were built, the ruins of which the traveller will meet with here and there in the central
portions of Florida, and which excite considerable
interest in the antiquarian. All of these institu-
tions were but outgrowths of the central establishment at St. Augustine.

The succeeding year the town was again

attacked, though, we should judge, quite accident-
ally. This time the Anglo-Saxons were the
aggressive party, for the first time venturing to
take a hand in this rather uncertain Florida
business. The hero was none other than that
restless and dauntless old sailor, Sir Francis Drake
of England. An expedition was fitted out by
some private adventurers, consisting of twenty-
six sail, with Drake as admiral. Their main
object could have been little else than plunder;
and they made no small fist of it either, for they
captured a number of Spanish seaports in the
West Indies, raked their commerce generally, and
wound up by paying a reconnoitering visit to the
coast of Florida, in the hope that they might add
to their spoils or their glory. They descried a
"look-out," or tower, off St. Augustine, and at
once made a descent upon the neighborhood.
Advancing, they encountered quite a formidable
fort, but poorly garrisoned, as the sequel showed.
On this they made a rapid advance, and were
greeted by a few random shots, when the guard
fled, leaving fourteen guns, and military stores,
including a treasury-chest of several thousand
pounds. This latter trifle was straightway escorted

with military honors to the fleet! The succeeding
day, Drake marched upon the town, which the inhabitants, with a very slight show of resistance,
evacuated, leaving it in the peaceable possession
of the English. History does not seem fully to
agree as to the burning of the place by Drake;
but since he did take the place, and as his hatred
of the Spanish was only second to their hatred of
him and his people, it is altogether probable that
he fired the town. Indeed, the weight of evidence inclines to this supposition. It is not unlikely, that, since they immediately left and re-em-
barked, but a small portion of the city was burned.
Drake had no intention of holding the country,
and, having completed his " swinging round the
circle " of Spanish settlements, set sail for home,
where he arrived in July of the year last above
named (1586).

The city of St. Augustine was, before this
attack, as large in point of population as it is to-
day. It soon recovered from this temporary, and,
as we believe, slight injury by Drake; and its
growth continued for nearly half a century,
though one writer asserts, that, about 1610,
the natives, from some unknown cause, made a

descent upon the town, capturing it, and utterly destroying it by fire. This we do not in any manner credit. In 1665, however, it was attacked and plundered by a Capt. Davis, an English buccaneer, — a class of persons who, from about Drake's time and for over two centuries afterwards, swarmed among all the islands of the West Indies and the adjacent seas. It was likewise about this period, that an English colony, of, more than likely, freebooters, established a settlement at St. Helena, on the St. Mary's River, within the then boundaries of the State, but now in Georgia.

In 1680, the Spanish governor became jealous of the great power of the chief of one of the largest and most powerful tribes in Florida, and decided to procure his arrest for *treason;* and it is not improbable that he had adjudged him worthy of death, even before a council had assembled to formally determine the question. He was, of course, pronounced guilty, notwithstanding his solemn protest that he was both loyal to and zealous in papal faith and power. He was sentenced to die ; and his last words were to his countrymen not to avenge his death, thus actually proving the sincerity of his asseverations. The tribe over

which he had held undisputed sway felt outraged
as they unquestionably were, in the death of their
beloved leader; and fierce and bitter war ensued,
resulting in the driving out of all the Spanish
settlers north of the St. John's River.   From this
period, the Spanish power in this colony commenced
rapidly to decline, while the English settlements
increased, till they had full possession of the
Carolinas.   The French had to a very great
extent withdrawn, and planted a colony at the
mouth of the Mississippi River.

In 1702, the English Gov. Moore, of South
Carolina, planned an attack on St. Augustine,
which proved disastrous only to himself.

Up to this period, most of the attacks on the
Spanish had been at St. Augustine, that being the
capital of their colony; but in 1703, Gov. Moore,
having a wish to retrieve his reputation, sought
and obtained a large force of Indians, accompanied
by a few whites, and made an incursion back from
the sea-coast on Lewis Fort, from which, by skillful maneuvering, he managed to draw out the
garrison, when a sanguinary battle ensued, resulting in the complete triumph of Moore's party.
The effect of this was to break the power of the

Spanish in Middle Florida; and they withdrew from that region, and the less reluctantly, since, in this conflict, their principal school and convents had been destroyed, together with their settlements. What had taken three-fourths of a century to build up had, as it were, in a day been utterly annihilated.

A few years of peace followed, — long enough to prove to the Indians, who had fought and won for the English their crowning victory in Florida, that "white men were very uncertain," and that those who were their friends to-day might be their enemies to-morrow. The truth of this they fully realized, as many better people have before, but more especially since these times of which we write.

They felt so outraged, that they did not hesitate to form an alliance with their ancient enemy, the Spaniards; and under their leadership, and in conjunction with them, they were marched into what is now Georgia, but at that period claimed as a portion of Florida, though in actual possession of the English, who had, during the last few years, added greatly to their numbers and wealth, with their chief settlement at Savannah. This combined army marched, about 1717, into Georgia,

4

only to succeed in obtaining a sound drubbing at the hands of the Britons and their Indian allies of that section.

The Spanish governor concluded, thereafter, to remain at home, and to quietly surrender the rule over a portion of his territory. The effect was, as the humorous Artemas Ward used to say, to make the ways of " Lo! the poor Indian," very hard indeed. The governor restricted their privi- leges, and drove them from their fields, so that they came to be kicked and cuffed in turn by the English or Spanish people, as fortune led them. Thus one of the most powerful tribes on the continent, the Yomasees, were, at this time, among the most abject. The Spanish now held St. Augustine and the adjacent settlements and territory on the Atlantic; while, on the Gulf coast, their corresponding strong point was St. Mark's.

The boundaries of Florida were now substantially diminished, and extended on the north between St. Mary's and the Altamaha Rivers, and westward by south to the Mississippi River. This western boundary was put in jeopardy by an attack on Pensacola, a place founded prior to 1690, and that had been held at times by the French and then by the Spaniards, up to 1722, when it was permanently relinquished to Spain by a treaty.

Difficulties were continually springing up between the encroaching English settlers on the north and the Spanish, culminating, in the spring of 1740, by Gov. Oglethorpe of Georgia planning an invasion of Florida, and particularly the reduction and capture of the city of St. Augustine. With an army of rising two thousand men (and Indians not taxed!) he marched on that place, seconded by an English fleet under Com. Price, who arrived off the harbor too late, however, to prevent re-enforcements reaching the town by the sea. An unsuccessful siege was prosecuted, chiefly from Anastasia Island, opposite the city. The distance was such as to render the shots ineffectual, though some of them struck and buried themselves in the sides of the fort. These shot-holes and abrasions may still be seen on the east and north-east side of the fortification, or could at the time of our visit. Oglethorpe was obliged to retreat, and fell in disgrace among his people : but later, in a counter-attack on his own colony by the enemy, he was, by some strategy,

enabled to regain their regard; and, in 1743, he
put on foot another expedition, which was marched
by land to St. Augustine. This proved more
disastrous, if possible, than the first. The Spaniards followed them on their retreat harassing
and killing many, aided, of course, by the savages,
who hung on their flanks, following and destroying
some of the English settlements in Georgia.
Finally, in 1748, a treaty was concluded between
Spain and England, which had the effect to produce a period of comparative quiet.

In 1763, Spain ceded Florida to Great Britain;
whereupon a large portion of the Spanish settlers
withdrew, and in their places came army-officers
and pensioners, together with emigrants from
Britain. The country soon became quiet, and
agriculture again flourished; but in 1783, after
twenty years of occupancy by the British, the
territory was re-ceded to Spain, and a corresponding hegira of the English occurred, as in the case
of the previous cession. It only needed this to
complete the desolation. Of course, the incoming
Spaniards did nothing, not knowing what moment
word would come of another transfer. Beyond a
single disturbing element of small proportions,

there was little to hinder the advancement of the peaceful arts ; but the country retrograded rapidly.

A more lamentable condition no state or territory could well find itself in. The industry of a century had been allowed to go to decay; and rapine and war had driven nearly every vestige of civilization from existence.

The United States, meanwhile, had grown — indeed were at this time — a powerful people, and looked upon this wilderness of Florida (for such it had now become) with eyes of longing, which was not wholly unnatural.

Troubles and disputes began to thrive along the northern borders, when President Monroe dispatched a small force in 1811 to act as a corps of observation ; but its commander, more zealous than prudent, soon found a pretext for occupying the whole of Amelia Island, on which the town of Fernandina is situated. His course was disapproved by the Government, and our forces were ordered to be withdrawn. While preparations for this object were going forward, the Spanish governor caused an attack to be made on the retiring forces, resulting in the killing of some of the troops. Instead of hastening, this act but detained our forces.

Again the commanding officer was superseded by
Gen. Pinckney. Meanwhile, the administration
sent two more commissioners to negotiate for a
transfer of the territory to the United States; but
this effort proved abortive. Internal strife and
commotion flourished; but, in the spring of 1813,
our troops evacuated the territory. The succeeding year witnessed an attempt on the part of the
British to occupy West Florida. The "Yanks"
needed but just such an attempt to give them
license to "move in;" which they speedily did,
under the direction of Gen. Jackson, five thousand
strong, and made an early call on the redcoats
who already occupied Pensacola. A trifling "un-
pleasantness" ensued: the English left on their
fleet which brought them; and Jackson proceeded
soon after to arrange for the more important duties
to come off at New Orleans on the succeeding,
and now memorable, 8th of January.

We pass the period which was well filled with
Indian conflicts and border warfare, to 1818, when
Jackson, in punishing the savages, was affronted
by the Spanish commander at Pensacola, in whose
vicinage he was. Whereupon he repeated the capture of the place, and sent the Spanish prisoners

and governor to Havana. This brought the question of a transfer of the territory plainly before the authorities of Spain. It was evident that "manifest destiny" would annex this territory to the Union, either with or without their consent. They at once entered into a treaty in 1819, which was ratified by an exchange of flags on the 17th of June, 1821.

CHAPTER IV.

THE transfer of authority over this peninsula was a notable event in its history. No human agency could well have averted such a result. Out of this fierce caldron of hates and retribution, order was to come, but not yet. Years of struggles and many precious lives were the real price we had to pay. No race of beings under the sun ever fought with such persistence and desperation as did these red men, the Seminoles, in the swamps of Florida. Their chief, Coa-cou-chee, was a brave and cunning savage, and his fighting-men excelled in every element of strength and agility. A pitched battle was out of the question. They were outnumbered, and fought, economizing life, behind the trees, or concealed from view in some jungle or thicket. A hissing arrow, and all else was silent. When least expected, they screamed their wild war-cry, struck, and vanished in the impenetrable forest. Nothing impeded their marches: they were expert as swimmers, and as tireless as the rivers whose dark, calm waters often swallowed and shielded them from their pursuers. Living on the fruits and fish which a provident nature abundantly supplied, they were enabled to pass rapidly over great distances, and seemed, indeed, almost ubiquitous. They were the terror of our troops; and considering themselves the rightful occupants and owners of the soil, and believing that this struggle was to be their last great effort, their courage and daring rose with the occasion. The history of this band of braves, so replete with thrilling interest and real romance, would afford ample material for a Walter Scott, from which to weave as many tales of this "place of flowers" as of his own bright and bonny land.

It is a sad history, that of the Florida aborigines. The rising sun brought with it no people

from the east who were to prove their friends.
Spaniards, Frenchmen, and Britons were all alike
in selfishness and duplicity.   Their lessons had
been severe and were well remembered : hence,
by the time the utilitarian population of the States
had acquired power and permanency, they, too,
were adepts in deceit and diplomacy.

History will account them a barbarous band;
but, while this is doubtless true, it should not lose
sight of the influences which wrought out this
result in their characters.   Would any people,
however enlightened, even in this age, escape the
doing of many acts of savage wantonness, if they,
as were these red men, were hunted from their
country, and their cities and public works leveled
to the ground, by some invading race of Titans,
who sought to change their lands back to their
primeval character and beauty, and who, by duplicity and right of might, defrauded them of even
their right to live !   That the Indians believed
themselves the children of the Great Spirit, whom
they sought to worship in their rude way, is true ;
and that they knew the white man to be an in-
truder is also true.

It may not be among the possibilities for the

two races to live together; and, if we concede
that it was in God's providence to establish the
white race on this continent, then we may, in
some degree, be reconciled to the march of events
here: but that can not palliate nor excuse the inhumanity and fraud continually practiced upon
these rude children of the forest, from the earliest intrusive visits of the whites to the present
time.

We have no right to judge of the former tribes
by the present degenerate fragments we see and
know upon our remote borders westward. They
are only the *residum* of a mighty race of men,
who for three hundred years have been plundered
and robbed of every possession, and schooled in
every vice, by the worst of our own people. It
was no crime to practice upon them; and, by systematic approaches, they have been entrapped into
every imaginable difficulty, for the sole purpose of
plundering them or the government who presumed
to shield them.

The problem of the two races living peaceably
near each other may never reach a desired solution; and, with the present roving bands, it may
indeed be impossible, in view of their long degra-

dation. Few venture now, since the effort of a
Schoolcraft has been unavailing, to even stand up
long in their defense. The shafts of sarcasm and
ill-concealed contempt, which is leveled at these
efforts, doubtless deters many whose hearts are
inclined to be moved in their behalf, so far at least
as to demand justice and fair treatment.

Gen. Jackson was too much of a fighting-man
to allow any high moral duties to enter into the
calculation of any of the rights of the Florida Indians; and the battles partook of the vengeful
spirit of the chiefs.

Osceola and Little Cloud were among the more
distinguished of the Indian chiefs; but of Coa-cou-
chee, though less is known, that little is full of
unusual interest. He was perhaps the royal
prince. He was a son of the famous King
Philip, and hence was an hereditary chief, which
was not the case with Osceola, who was an
accidental or elected chief, and only a half-blood
Indian. Our royal scion might have served for a
model,—tall, straight, and slender, with the eye
of an eagle, and as shrewd, agile, and untiring
as history has painted him: yet he was captured,
but subsequently escaped by leaping down from

the tower of San Marco Castle at St. Augustine, where he had been kept imprisoned. He was again captured, and this time so heavily ironed and so long imprisoned, that his proud spirit was bent if not broken ; and, to procure his release, he was compelled to sign articles of removal to a home west of the Mississippi River, and also to induce his tribe, now but a remnant, to go with him. This they consented to do for their chief's sake. Coa-cou-chee, believing in the protection of the Great Spirit, was measurably reconciled to the loss of his beautiful home and glorious hunting-grounds, and to an exile in what was to him an unknown and distant land. The historian Sprague tells the story of a dream as related to him by this Prince of the flowery land, which occurred while confined in the dungeons of the fortress at St. Augustine.

" The day and manner of my death," he says, " are given out, so that, whatever I may encounter, I fear nothing. The spirits of the Seminoles protect me ; and the spirit of my twin-sister, who died many years ago, watches over me. When I am laid in the earth, I shall go to live with her. She died suddenly. I was out hunting ; and, when

scated by my camp-fire alone, I heard a strange
noise, — a voice that told me to go to her. The
camp was some distance off; but I took my wife
and started. The night was dark and gloomy;
the wolves howled about me. As I went from
hummock to hummock, sounds came often to my
ear: I thought she was speaking to me. At day-
light I reached the camp, and she was dead! I
sat down alone, and under the long gray moss of
the trees, when I heard strange sounds again. I
felt myself moving, and went alone into a new
country, where all was bright and beautiful. I
saw clear-water ponds, rivers, and prairies upon
which the sun never set. All was green: the
grass grew high, and the deer stood in the midst
looking at me. I then saw a small white cloud
approaching; and, when just before me, out of it
came my twin-sister, dressed in white, and covered
with bright silver ornaments: her long black hair,
which I had often braided, fell down upon her
back. She clasped me around the neck and said,
' Coa-cou-chee! Coa-cou-chee!' I shook with
fear. I knew her voice, but could not speak.
With one hand, she gave me a string of white
beads: in the other, she held a cup sparkling with

pure water. As I drank, she sang the peace-song of the Seminoles, and danced around me. She had silver bells on her feet, which made a loud, sweet noise. Taking from her bosom something, she laid it before me, when a bright blaze streamed above us. She took me by the hand, and said, ' All is peace ! ' I wanted to ask for others ; but she shook her head, stepped into the cloud, and was gone. All was silent. I felt myself sinking, until I reached the earth, when I met my brother Chilka. He had been seeking me, and was alarmed at my absence."

Coa-cou-chee was undoubtedly greatly influenced to sign the articles of removal in view of this dream. There are, perhaps, no people so studious as the untutored savage to follow what they believe to be the dictates of the Great Spirit, whom they hold in such supreme reverence and awe.

When he was asked what had become of the " white beads," he declared he had lost them in the darkness of his dungeon at St. Augustine. We can the more readily credit this, having ourselves been permitted to enter them for inspection ; and surely darker and more dismal caverns of prisons can not possibly be even imagined. To say that they were constructed under the supervision of the Spaniards is sufficient assurance of their completeness for every purpose for which they were made.

Standing once more free, on the deck of the ship in Tampa Bay which was to transport him and his people beyond the Mississippi, folding his arms upon his breast, he gazed intently upon his native land, wrapt in silent contemplation of the past and its eventful history; and, running in thought up the long line of his royal ancestry and back to his own and his nation's humiliation, his proud heart for the first time melted, and blinding tears veiled his sight. Turning silently away from view of the land which held all that was dear to him, he breathed a brief prayer to the Great Spirit; then, gathering up his mantle, sank down upon the deck, never again to gladden his eyes with the scenes of his youth he loved so well.

The Union had acquired, by the cession of this territory, 59,268 square miles, embracing about thirty-eight millions of acres, or an area nearly equal in extent to all of New England. A very large proportion of this is an extensive marsh, which during the rainy season, between June and October, effectually prevents an overland transit from one shore to the other in the southern portion. The northern and eastern section of the peninsula is of a level character; the middle and western portion will be found in places quite uneven and considerably elevated, yet no mountains will be found.

––––––––

The geological formation of Florida is of comparatively recent origin, and is indeed among the latest-formed land on the continent, excepting only the delta of the Mississippi River. Its present condition would, even to the ordinary scientist, give many proofs of this; but we are not without the opinions of learned *savans* on this question, who hold that the whole peninsula has been formed by the successive growths of coral reefs, added concentrically, from north to south, to the first deposits, while the accumulation between these reefs has been a mixture of coral and fragments of shells: the coral prevailing in some parts, as in the regions of the everglades; and in other

5

portions, especially the northern and eastern, the shell. In various places, imbedded in these shells, fossil remains of animals now inhabiting the country may be found. Upon this marine limestone formation and its inequalities, fresh-water lakes exist. On the shores of one of these, — Lake Monroe, — near the head waters of the St. John's, and through which that river passes, we exhumed the remains of human beings. They were in a most remarkable state of preservation, and indicated a race of men equal in stature to that of the strongest of Anglo-Saxons. It is not known how long a period these remains have been buried; but a learned writer has said that Upper Florida, as far south as the head waters of the St. John's, constituted a prominent peninsula before Lake Okee-cho-lee was formed, and that the whole of the southern extremity of the State, with the everglades, has been added to that part of the continent since the basin has been in existence in which the strata with human bones have been accumulating. Agassiz, in referring to the formation of the lower and later half of the peninsula, that part which lies south of the fresh-water system, — being in extent some three degrees of latitude, — assumes, that if the rate of growth be one foot in a century from a depth of seventy-five feet, and that each successive reef has added ten miles of extent southward, it would have required, on this computation, one hundred and thirty-five thousand years to have formed the southern half of the peninsula above referred to. The peculiar growth of some of the islands of the Carribean Sea is quite like that of the lower half of the peninsula of Florida. They are formed on the tops of some mountain under the sea; and, through the agency of animals which inhabit the water, these coral growths are established on them, and are gradually elevated to the surface.

It is altogether probable, that, previous to the last four centuries, there was nothing known of this continent by the Europeans. After its discovery, the appellation of the New World was given to contradistinguish it from the eastern hemisphere. It was to them indeed a new empire; and its marvelous attractions in climate, fruits, and flowers led them to great exertions in the way of discovery and exploration. They little dreamed that this vast continent was peopled throughout its entire extent, and that its birth antedated that of

their own " Old World ; " yet later investigations
prove this to be true, and also startle us of the
present century, when we are informed that an
exuberant flora existed in Louisiana more than
one hundred and fifty thousand years ago, and that
in all probability the human race existed in the
delta of the Mississippi over fifty thousand years
ago. The reasons by which these conclusions are
reached are generally well known and accepted
by the more eminent geologists.

In the course of excavating for some public
works, it is stated, at New Orleans, the laborers
came to and cut through four successive growths
of cypress timber, the lowest so old as to cut like
cheese ; and one tree was found full ten feet in
diameter. The growth of these trees is very slow,
requiring about one hundred rings to give an
inch in diameter from the center. This would
make the age of the largest about six thousand
years. It is not known how many generations of
these trees have perished on each of the several
levels ; but it is assumed that at least two have.
This would give to each strata of cypress growth
twelve thousand years ; and the four growths mul-
tiplied by this, to which must likewise be added

the era of plants prior to the cypress, estimated at fifteen hundred years, also the present live-oak platform, or era, thought to be about the same in time, would make a grand total of fifty-one thousand years. And it was *under* this *fourth* growth of cypress, that a skeleton of a human being was found, belonging to the aboriginal American family ! This was found within about fifty feet of the surface, while the total depth of the *delta* itself is between five and six hundred feet, and is believed to have required fully one hundred and fifty thousand years for its accumulation.* This is probably somewhere near the age of the lower half of the peninsula of Florida. How much older the upper portion may be is unknown; but that even dwindles into comparative insignificance when we contemplate the age of the elevated portions of the continent. This much as to the formation and age of the peninsula; and we pass to the consideration of its admission into the Union.

The territory, embracing the two provinces of East and West Florida, was consolidated, and

* See Types of Mankind.

Gen. Jackson was its first territorial governor. In 1845, Florida was admitted into the Union ; and for a time there was comparative quiet, and new settlements were established. But poor, harassed Florida was not long to enjoy either peace or prosperity. The secession movement in 1860 raged in all portions of the South ; and this State, with others, was by fraud and violence carried out of the Union. It soon became again, for almost the hundredth time, a theater for war. The general government at once occupied all of its chief seaports, and places along its rivers, with troops ; and incursions into the interior were the standing order. No great battle was fought within its borders, though numerous and sanguinary conflicts were caused by the various raiding-parties in all the more northern portions of the State. Finally the war ended, and peace again has spread her wings over this oft devastated and many times ruined people. And now let us hope that this is the inauguration of a prolonged and flourishing period which time has in store for her.

Thus far we have considered Florida as a unit : while giving its general history, we could scarce have pursued any other course, and make ourselves at all intelligible to the reader. It was perhaps desirable to cover the whole ground in that particular, as a clearer idea would be formed of the eastern half, as related to the State and its eventful history. The territory was, however, early divided into two provinces, known as East and West Florida; and in each there was for many years a local head. As it suits our present plan to continue the division, we shall hereafter speak mainly of the eastern half, since our travels and observations have been confined to that region, except, perhaps, in a casual way, when we come to treat of the climate and the products of the State in future chapters.

CHAPTER V.

IT was five o'clock in the morning, when the steamer "City Point" rounded the southern end of Amelia Island, and steamed up to the wharf at Fernandina. We roused ourselves from a broken, weary slumber, and, drawing aside the blinds to our state-room window, cast our eyes for the first time upon this antiquated town, situate near the mouth of St. Mary's River, and at the extreme north-eastern corner of the State. It was low tide; and the wharf had a kind of black-spider look, which did not charm, neither did it deter us from dressing, and proceeding on shore. The wharf is a projecting one, with a

72

commodious warehouse at the landing. The morning was darkened by a sea-fog, which now commenced to drift in, partially obscuring observation. A brief excursion satisfied us that this was an uninviting place to stay in. Its look of decay and hard times seemed as legible as if written on every plank in town. Beyond the sight of the railroad, which runs from here to Cedar Keys, there was nothing of special interest. Hurrying back to the steamer, which till now had seemed never so friendly, we were glad to receive its protection, and quite rejoiced when the wheels splashed the water, bearing us away. The truth was, we were disappointed; and this made us feel homesick, and created a wish that the steamer would head northward, and away from the desolate shores of Amelia Island.

A few hours of steaming brought us near to the bar in front of the mouth of the River St. John's, — called in the times of Bartram, St. Juan. The fog was dense, and the locality dangerous for an advance : thirty minutes of this bewilderment, and the penalty paid is detention till the succeeding tide. - A few moments, however, and a whiff of air drives the fog, and lifts it skywards. The

pilot is proved skillful, for we lay but a few yards
from the mouth of the river. Crowding on all
steam, we barely pass over the bar; and then,
fair sailing. And this is the famous river of
which we have so much heard, and whose praises
are being sounded far and wide by every tourist
who visits it.

Passing the headlands, an open flat, or savanna,
appears, in width perhaps five miles, extending
however, many miles up along the coast as
far as the eye can reach. The river crosses, in a
westerly direction, this low tract, much resem-
bling the Jersey flats between Newark and New
York. We soon pass on, and by the solid
banks of the main land. Three hundred years
ago and more, Jean Ribault threaded his way
over the dangerous quicksands at the entrance of
this noble river, and, proceeding up, landed, and
built a monument of discovery; then sailed away,
to return in a few years with a colony of his
countrymen, who were subsequently slaughtered
by the Spaniards, as we have seen in a previous
chapter.

Not more than a mile from the mouth of the
river, and on the banks of this low marsh, we

saw, to our great wonderment, as the steamer hugged the banks, *seven* alligators of varying sizes, and all within a few yards of each other. They did not stir, or even wink, so far as one could observe, though the steamer's waves washed up to and partially over them. This was a new and curious sight, and gave promise of some fine sport, which it was hinted we should have on the upper portion of the river, where they abound.

Having passed through this savanna, or low marsh, we steam close to the bank of the river, keeping the principal channel; and, as we stand upon the decks, our attention is called to the first landing, May Port, an inconsiderable place. A rickety old wharf of palmettoes, with several corresponding habitations, constitutes every thing. But here we catch the first glimpses of a tropical growth. A few rods distant from the landing stands a small grove of palmettoes; though not over twenty-five feet in height, yet their peculiar character attracts the attention of the Northern traveler. Casting your eye farther inland, and the green forests tell of a perpetual spring; and we now realize that we have passed from the chilling winds of a Northern winter to the genial skies of that sunny land, of which we so often hear, but seldom or never experience in any other of the Southern States.

A recent traveler, entering the river, and sailing westward across the marsh before referred to, to the point of the main land where the river turns abruptly southward, asserts, that, in making this turn, a curious atmospheric phenomenon is met with : a current of warm air is so suddenly encountered, drawing down the river, that the change is like going from a cold hall into a warm room ; that this change can be observed, not only on a calm day, but when a high wind is blowing. It is undoubtedly the gate, as it were, to that tropical land where perpetual summer reigns.

The invalid and the tourist flock eagerly to the steamer's decks, to drink in this balmy air, and feast their eyes on the green foliage of the forest which flanks on either side the water's edge. The drooping spirits of the passengers rise as the steamer advances up the broad and beautiful river which so delights all eyes and gives eloquence to every tongue. The surface of the country is low and uniform, presenting none of those grand and startling features which characterize the Hudson or render the Connecticut surprisingly lovely; yet it possesses a marked and pleasing contrast to the muddy waters of the Santee, Cooper, Savannah, and other rivers of the South. Its shores of glistening white sand, and rich verdure of Southern foliage along its banks, with its broad, expansive, slow-moving waters, attract and interest all, and especially those who are accustomed only to the bolder outlines and scantier vegetation of the more northern water-courses.

The course of the River St. John is noteworthy. There are but few rivers on the globe, of any magnitude, that run in the same direction. Rising in the southern half of the State, fed by and flowing out of that vast fresh-water basin formed by the everglades and savannas, and commencing below the 28th degree of latitude, it runs in a due northerly course for about two hundred miles, when it abruptly turns eastward, and empties into the ocean. It is a magnificent highway, and is to Eastern Florida what the Hudson is to Eastern New York. We know of but two rivers of any importance whose direction is exactly the same. One of these drains the

northern limits of British America, and debouches
into the Arctic Ocean ; and the other, having its
sources in Middle Europe, into the Baltic Sea.
We have said the St. John's was a noble river: and
it is truly so ; for what river can the reader call to
mind, which he has seen, that for over one hundred miles of its length has a general average of
even one mile and a half? We know of but one
other in this country, which would come even up to
this estimate. Yet shall the truth be credited,
when we state that this river will average about
three miles in width for the distance above named!
In places it expands to twice that; and we can not
recall seeing it contract at any point to less than a
mile, after you get fairly into the river, until Lake
George is passed, above Palatka, when it narrows
suddenly, and continues thus to Lake Monroe,
which is the head of regular steamboat navigation,
though a small steamer does go on as far as Lake
Harney, and even beyond, but only at the instance
of excursion parties. This portion of the river,
known as the " Upper St. John's," we shall speak
of in a subsequent chapter ; meanwhile we return
to " The City Point," on which we were just now
steaming up, on our way to Jacksonville, some

twenty miles from the river's mouth. Passing
Yellow Bluff, a small landing of trifling importance, at which the steamer halted, and rounding
a bold point in the river, we heave in sight of
Jacksonville, the commercial metropolis and business center of Eastern Florida. All eyes are
upon the town, and all are pleasantly surprised to
see what is apparently and actually a flourishing
little city. Its bright, cheerful appearance, having
a fine view, as we near the pier, is attractive, and
so differing from Fernandina, the first point
touched in the State, that we already are having
our original anticipations more than realized, and
begin to rejoice that we have made the journey,
which was undertaken at the outset with some
misgivings.

JACKSONVILLE

is situated on the western bank of the river. There
is a ferry across to the opposite shore, where are a
number of pleasant homes, and among them is that
of the present State Executive, Gov. Reed. The old
stage-road, called the " Kings Road," runs from this
place along the coast, north, into Georgia, and across
the river south-west some forty miles, to St. Augustine. The travel now goes to that place chiefly by the

way of Picolata, some thirty miles up the river, thence by stage eighteen miles across. The railroad connections of Jacksonville are considerable, and give a life and bustle to the town which we scarce expected. True, but one line of road enters the city; but this runs in direct connection with the various roads in, and those entering the State. At Baldwin, some twenty miles west, connection is made with the Fernandina and Cedar Keyes Road, which crosses at that point. Proceeding westward to Live-Oak Junction, connection is made with the Florida and Atlantic and Gulf R. R., for Savannah, which is distant from Jacksonville twelve hours and a half by the night express-train. Continuing westward from Live Oak, the road runs through to Tallahassee and St. Marks on the Gulf.

Jacksonville, named for Gen. Jackson, is an incorporated city, having the usual city officers and a board of trade. It is regularly laid out, having eight principal streets about two miles long, and running parallel with the river: these are crossed by others at right angles, their length being easily extended, *ad infinitum*. Shade-trees abound, and of fine size, consisting of live and water oaks, the latter making for this purpose much the finest tree. Its luxuriant and rapid growth and hardiness have caused it to be adopted. They are both evergreens. From the river, northward a short distance, the land has a gentle elevation; then descends, terminating in a low tract or swamp. On the north-western side of the city is a beautiful bluff, covered with attractive residences, shaded with stately trees, and bearing the pleasant name of " La Villa." The view up the river from this point is delightful. Fifteen miles away is Mandarin, a projecting tongue of land, which, from where we stand, seems to sever the river. The river has such a reach of view from this point, and such breadth of waters, as to give the appearance of a lake with wooded shores, fifteen miles long and from three to five miles wide. Passing on from this village a short distance, and toward the river-banks, you come to the new town, appropriately named " Riverside," and likewise a suburb of Jacksonville. This place is now being built up, and designed to be made a select quarter. Its streets are spacious, and a park has been reserved, and many shade-trees planted. None but fine buildings are to be erected; and all must be

6

painted, instead of whitewashed, — a practice followed by owners of cheaper dwellings in the
city.

No city of the South can show better evidences
of prosperity, or a larger increase of population,
relatively, than Jacksonville.

The waste and stagnation caused by the war
has in this place quite disappeared.

On every hand, we behold the magic touch of
Northern hands and Northern capital. Whichever
way we turn, new buildings and stores greet our
view; old ones being enlarged, streets cleaned,
and substantial, or at least convenient plank-
walks laid where lately they were innocent of any
such improvement. On the river, wharves are
enlarged, till now its accommodations seem ample
for an extensive commerce. In short, new im-
pulses and new ideas have seized the town; and its
present watchword is "Forward."

The population of the place is about six thousand, increased during the winter months to about
eight thousand, by the influx of strangers from
all portions of the North, seeking temporary
homes. Not until the past year has there been
any thing like adequate accommodations for this

large number, not taking into account the constant flow of visitors going and returning. Great exertions have been and are being made to supply the demand for small houses, and Northern energy and capital are kept busy. Even that venerable and itinerant agriculturist of " The Tribune," Mr. Solon Robinson, is busied almost night and day, we should judge, in superintending and building houses of this class.

As to hotels, there are several; though the most pretentious is " The St. James," erected in time to reap considerable of a harvest from last season's business. It is a large, overgrown, yet rather imposing establishment, capable of receiving two hundred guests. It is situated about half a mile from the steamboat landing, and likewise the railroad depot, and seems every way adapted to the wants of the traveling public. Another house of equal capacity is, we learn, contemplated. " The Stickney House " has a good reputation, and several friends stopping there spoke very well of it. There are, in addition, the " Price " and " Tyler " houses, and numerous private boarding-houses kept by Northern people, with pleasant surroundings, and quite as inviting as the hotels to those seeking but a temporary home. Price of board ranges from twenty-five dollars to ten dollars per week at these various places. Unfurnished cottages can be had from twenty dollars to thirty dollars per month. Pleasant furnished rooms in private houses, including fuel, light, and attendance, are from four dollars to six dollars per week; and the best board, without rooms, at the hotels, is eleven dollars per week, while in boarding-houses it can be obtained at a less figure. Laundry work is extra, and costs usually one dollar per dozen.

The town contains ten churches, four of which are colored, the most of them having pastors. But the most of them need paint, with a little upholstering, which, with a few additional yards of carpeting, would, even in Jacksonville, — more favored than many towns of the South, — add much to the comfort of both residents and strangers. But time will change this, and we must not expect overmuch from those who are already doing great credit to themselves.

The colored element in Jacksonville is numerically the largest. They support churches and schools, and are represented in the city government and in the courts, and are to be found, as they always have been more or less, in the mechanics' shops and behind the counters of some stores. Under and by the direction of the Freedman's Bureau, a normal-school building was some time since erected, at a cost of fourteen thousand dollars, which offers superior educational advantages for all classes ; though we are informed that none but colored children avail themselves of its benefits.

The chief article of manufacture and export, not only from this city, but from *Eastern* Florida, is pine lumber. Jacksonville has over a half-dozen extensive steam saw-mills, kept constantly in motion, cutting up logs into lumber for the markets north. Whole forests go down to keep the maws of these monsters supplied. Whether this will always last is doubtful ; for, in the aggregate, a small army is kept plying the axe just back of the main river and along its tributaries. We say "just back " from the river. Along the banks of the watercourses no pine lumber grows ; but a short distance back from them, in many cases less than a hundred rods, pine lumber abounds : indeed, it is the chief thing in this State, as it is in all the States belting the Atlantic, as far up as the James River in Virginia.

The commerce of this place is equal to that of some cities in the North of three times its population. Brigs, schooners, and steamers, several in the aggregate, arrive and depart almost daily for different ports up the river or along the sea-coast from Savannah to Portland.

A daily mail arrives by rail from the North; and this, with the telegraph, gives Jacksonville an importance which no other town in East Florida possesses: indeed, this is the only town which has both of these *luxuries*, though all the river-towns have abundant mail facilities. There is a telegraph to St. Augustine.

In our next chapter, we shall continue our voyage up the river, noting every thing of interest which came under our observation.

CHAPTER VI.

FROM JACKSONVILLE TO GREEN-COVE SPRINGS.

A BEAUTIFUL day is this in which to resume our trip. The sun comes out full and warm : overcoats are not thought of. It is quite like a warm day in May, when the bees are busy with the flowers, and apple-orchards are more fragrant and delightful than any Eastern perfume. The air seems redolent with the odor of balsam and wild flowers, wafted thitherward by southern gales, while on either shore the foliage is bursting into freshest and tenderest green, contrasting finely with the dark verdure of the perennial laurel, oak, and magnolia, which line the river-sides.

It is a common term, " up the river," or " up

north ; " and since most rivers run southerly, up
the river gives an impression of going northward.
We say " up north " and " down south ; " these
terms are all understood when thus used : but to
say " up the river," in speaking of the St. Johns's,
means *going south*, since the river runs from the
south toward the north.   Hence, every revolution
of the steamer's wheels carries us " up the river,"
and at the same time " down south."

By the time our steamer had gotten fairly under
way, the passengers had chiefly assembled upon
the decks to feast their eyes on each interesting
point and feature of the river, and to gaze in wonder at its emerald shores, which now stand several
miles away on either hand, while our vessel plows
its trackless way in the center of the river.

Turning toward Jacksonville, it seems but a
small village, and nearly hidden at that by its
luxuriant shade-trees.   If one were suddenly placed
where we now are, the first query would likely be,
" What lake is this ? " and such it seems to those
who have been accustomed only to the streams
and rivers of New England.

Our passengers may be divided into four classes,
and preponderate in the order in which they are

named : first, those who are seeking pleasure ; second, those who are seeking health ; third, those who come to invest in land for speculation or occupancy ; fourth, the resident population living along the river, and traveling on their ordinary business. The first and more numerous class appear generally to realize their object, and are greatly interested and delighted with every thing they see, which is so novel, if not so wonderful, as to engage and enliven them constantly. It is indeed a source of real amusement to observe the younger portion of this class rush from point to point, as some stray member of the company spies some near or distant object thought worthy of the others' attention. The next class, the invalids, are not so demonstrative, but show many signs of confident hope and real delight to find themselves in a climate where they can, with comparative impunity, not only sit and walk in the open, genial air without detriment, but with real benefit to their spirits and general health. This class is largely composed of those on whom disease has so recently fastened itself as to scarce indicate its presence. These must be very greatly benefited, if not wholly cured ; while others are so far reduced as to make traveling of any kind a

burthen and labor. Such we greatly pity; for in
their faces hope is a stranger, and they can not
long remain in any clime or earthly habitation.
They come here to die, far from home, deprived
of the sympathy and attention of loving and watch-
ful friends, and of home-comforts which money
can not purchase. It is true that the people here
are kind, and give such attention as is in their
power; yet this does not compensate for a wife's,
a mother's, or a sister's tender interest and presence.

The speculative class may be known by their
inquisitive look. Every feature of their face seems
an interrogation point. They seem bent on ex-
tracting from you information as to the best location, kind of soil, and climate, and if there are any
snakes in that neighborhood, and if we know of a
good place for a *hotel!* Many are the times we
have been greatly amused by the ingenious
methods of these persons in their attempts at elicit-
ing their prized information. To such as we felt
were anxious to locate, our limited knowledge was
freely imparted; but to the other portion of this
class, those looking for a " good thing to hold,"
our snake-stories had no relish: they were twice too

long and plenty for their fancy, and they invariably "moved on," quite astonished and disappointed.
Nothing but the "pesky rebels and them are
snakes" gave them cause for any apprehension.

There is a good deal of a fitful fever of speculation, which is prevented from settling into any
thing like a decided type, in consequence of the
difficulties surrounding a good title.   Then, too,
many of those who hold lands are women (widows,
made so perhaps by the war); and these especially,
if asked to set a price, fear to do so lest they should
regret not having asked twice as much.   Pur-
chasers are not a little disgusted, and generally
return without having made an investment.   By
and by these things will change; and, even as it is,
new settlers are working their way into all portions
of the State.

A long shrill whistle by the steamer an-
nounces a landing, and the wharf at

MANDARIN

lies just before us. This place is famous as the winter
residence of Mrs. Stowe.   It is but a small settlement, of perhaps a half-dozen dwellings scattered
along a tongue of land projecting toward the center of the river, and forming a kind of bay.   The
outlook is northerly toward Jacksonville, which

lies half a score of miles distant down the river. The location of Mandarin is exceedingly pleasant, and the view fine and far-reaching.

The water near the bank of the river on the Lower St. John's is quite shallow, necessitating a projecting pier of some length at all of the landings ; and this is especially the case at Mandarin. Mrs. Stowe's house is near the bank, and but a few rods to the left of the shore-end of the pier. It is of a dark brown color, of very moderate cottage size, wholly unpretending in appearance, and quite inexpensive. The chief feature of her place, as seen from the river, is its magnificent towering shade-trees, — the water-oak. They attract and fasten the eyes of all by their unusual size and beauty. Their wide-spread and over-shadowing branches give an air of seclusion and dignity to the quiet home beneath. Her place, we are informed, consists of some forty acres in the rear of her dwelling, with three or four acres of orange-trees, large enough to bear fruit. This grove is being added to, and doubtless in time she will possess a very large and beautiful orchard and one of great value.

There is but one small boarding-house at Mandarin, and few disembark ; none that we notice, except such as reside in the vicinity.   The mails being exchanged and freight left, we steam on, and up the river; and, rounding the point,

HIBERNIA

is seen just ahead, on our starboard bow.   Beyond is a spacious, overgrown New-England-appearing farm-house, hid behind clustering trees and branches, on a pleasant and inviting spot of land; and, aside from this, nothing else appears.   The steamer stops, and here a goodly number leave for the hotel.   Here and there appears an occasional fine plantation, with comfortable dwellings and numerous outbuildings, once the house of the lordly planter and his slaves, but now simply monuments of a past and exploded system. Next is

MAGNOLIA,

with a large and commodious hotel.   This place was formerly owned by Dr. W. D. Benedict, and originally selected by him after careful examina- tion for its attractive natural scenery and high sanitary reputation.   This property has recently been purchased by Dr. S. Rogers and O. F. Harris, formerly of Worcester, Mass., and by them greatly improved; the buildings having been en-larged, thoroughly repaired, and newly furnished throughout for the accommodation of about one hundred guests. From the reputation of these new proprietors, we should judge visitors would find this a desirable winter home. The house is to be opened by the first of November. Its proximity to Jacksonville and Green Cove Springs give it some advantages not possessed by towns further up the river. The facilities for reaching these several towns, situate on the banks of the St. John's, are the same. The Savannah, Charleston, and river steamers all touch, going up and down, long enough to leave or take aboard both passengers and freight. The captain of one of the river steamers states that Magnolia Point, con-tiguous to Magnolia, is one of the highest, if not the highest point of land extending into the river between Jacksonville and Palatka. It certainly has two commanding views: that down the river extends to Mandarin, and up the river your prospect is not less circumscribed. Black Creek, a navigable stream for twenty miles, and quite well settled, empties into the St. John's but a short distance north of Magnolia Point. Small steamers make trips regularly once a week from Jacksonville up Black Creek as far as Middleburg, and not unfrequently oftener, as parties or business require.

Immense quantities of pine timber and lumber are floated down this creek to Jacksonville ; and an equal quantity of alligators might be taken out, we have no doubt, if there were only as good a market for them as for the lumber, which at present unfortunately does not appear. These animals, after passing the first few miles at the mouth of the river, — which is through a low marsh, as we have stated, — are scarcely seen between Jacksonville and Palatka, not because they are not numerous enough, perhaps, but for the reason that they find more quiet and better feeding-places in the coves and up the large and small creeks which flow into the St. John's ; and, among these, Black Creek seems a favorite resort.

Two miles beyond Magnolia, past the first point of land, a beautiful spacious bay is seen, on the shores of which stands

GREEN-COVE SPRINGS.

This is the first *village* after Jacksonville, and the only considerable place between it and Palatka, a distance of one hundred miles (round numbers). More than a score of well-dressed people are standing on the end of the pier, awaiting our arrival : and full that number land with piles of baggage, looking quite as though they had come to stay ; and this is likely, for here the tourist and invalid find extra attractions as the name would indicate. There is a healthful and charming spring here, where all may bathe and frequent, without money or price. The village is scattered along and back from the banks, in a somewhat straggling manner, with not over thirty dwellings ; yet you feel their influence and companionship, and this drives off the blues, and is certainly an advantage to the invalid as also the traveler.

The accommodations for guests are quite ample ; though, if they were greater, they would not stand idle. We know of quite a number, who, coming in on the steamer, were informed that all rooms were filled, and were obliged to re-embark and seek some less frequented place. There is a large

hotel here, The Union House, kept by Mrs. Eaton,
though nominally in charge of Mr. Remington, a
Northern man.   This house will receive comfort-
ably about fifty guests.   It is large and airy, but
shockingly needs a coat of paint and some repairs,
and also more prompt attention to guests, to make
it more welcome as a home.   Still, it does very
nicely; and doubtless it is the intention of the
owner to introduce the suggested improvements
by another season.   There is also a large boarding-house near the spring, whose proprietor has
such an odd foreign name that it quite escapes us.
His house accommodates twenty.*   Then a few private families open their houses and receive guests.
Capt. Henry Henderson and his warm-hearted
companion never turn any away if they have any
sort of accommodations for them.   Their house is
not large, but it covers a great many people.
Capt. Henderson is a favorite with the visitors.
His extensive acquaintance with, and long experience in the State, have given him a vast fund of
information and anecdote with which he beguiles
the evening hours,— the center of attentive circles.
Everybody knows him among the natives; and he

* Captain Glinskie.

7

is much respected and esteemed by them as a man
of spotless integrity and honor. He is a native of
St. Lawrence County, New York, and has been
in Florida forty-five years: he is now seventy-
three years of age, yet he boasts of his ability to
" take his horse and hunt with any of the boys."
He enjoys a social game of eucher; and scarce an
evening, when not interesting some one with his
wild and fascinating adventures, he might be
found seated with Tompkins of Stony Point, Mer-
rill of Vermont, and Hughes of Ohio, each trying
to outdo the other in friendly strife. The captain
is hale and hearty still.

Our home in this village was with the family of
P. R. Coleman of Orange County, New York.
He has been in the State but a couple of years.
He came out alone, afflicted with chronic rheumatism; but settling at this point, by systematic
bathing and copious drinking of the pleasant waters
of the spring here, he claims to be entirely cured:
certainly he has now a fresh and vigorous appear-
ence. Deciding to settle here, he sent for his
family; and with them we found a comfortable
home. Mrs. Coleman, a most amiable lady,
brought with her what many people can never

carry, for obvious reasons, the perfect art of housekeeping as known in our best Northern homes.
Should this family remain at Green Cove, we feel
bound to assure all who are so fortunate as to find
lodgings under their roof, that they need borrow
no trouble or anxiety as to their comfort while they
remain.

The price of board varies from ten to fifteen
dollars per week. By the day, at the hotel, it is
three dollars; at the private houses, two dollars.

Fresh fish, venison, and fowl grace the tables in
this village, and can usually be had in abundance.

The river is, in fact, full of every kind of fish,
and on its surface float nearly all kinds of waterfowl; while the woods afford an abundance of
game and wild animals, of which we shall more
fully treat in some succeeding chapter.

The spring is the attraction at Green Cove,
distant not over ten rods from the river, and
double that from the principal hotel. It is owned,
as are most of the village lots, by two parties, —
Mrs. Ferris and a Mr. Palmer at Jacksonville;
who being opposed to improvements, and seem-
ingly to every enterprise as well as to their own
interests, suffer the spring to remain surrounded

by a clump of wild trees, which, however, serves
a good purpose in screening bathers ; yet it is not
to the interest of the villagers to have affairs continue in their present unimproved state. The
spring has scooped for itself a bed, twenty by
fifty feet, and lies some ten feet below the surrounding level. It flows freely from its fountain-
head into this spacious reservoir, of an average of
five feet in depth, and then runs rapidly off to the
river. The water is at a temperature of, we
should judge, about seventy-five degrees, very
pleasant, and thoroughly unharmful to drink. It
bursts up with considerable force, and clear as
crystal. It has a slight sulphurous taste, and leaves
slight traces of the same mineral on the sides of
the spring. Both magnesia and iron are, we
think, held in solution. Every one spoke in high
praise of this spring-water, both as a drink and for
bathing uses ; for the latter of which, regular
hours are assigned to the different sexes. A dozen
or more rude yet convenient dressing-houses sur-
round it.

In a short time, Green Cove must become, for
invalids at least, the chief point of attraction on the
River St. John's. This conclusion we have come

to after having visited every other place on its entire length. The sportsman, however, should locate on the Upper St. John's, many miles above this. The tourist and pleasure-seeker will not unlikely visit all the towns along the river; and even this latter class would find this point a proper and convenient one as a head-quarters. It has quick and frequent communication with Jacksonville, often several times each day; the mails are very regular; and, in case of any exigency, the telegraph is accessible at the latter place, where also the usual accessories to comfortable housekeeping may be easily obtained.

Up Black Creek to Middleburg, the head of navigation, is some twenty miles. The county-seat is located there; and before the war it was a respectable village, with a convenient hotel, where numerous invalids found cheap and not uncomfortable quarters. This, with other buildings, was in one of the many raids through this section destroyed. A small steamer runs up there once each week from Jacksonville, when not otherwise employed; which happened to be the case at the time we were ready to go! A trip across the country through the woods was resolved upon.

The needed equipments were a horse and " Colt,"
the latter of the revolving breed; to procure which,
we were directed down the river road, a mile away,
and thither we bent our steps. The first half-
mile we had no difficulty in following the " road; "
after that, it became a serious question : but, keeping the edge of the river, we came where several
old slabs, and a log or two half submerged, lay
across the arm of a projecting pool, overgrown
with pond-lilies, and filled with water as black as
ink, colored doubtless by the decaying vegetable
matter it held. We essayed to cross. Rapid motion was a necessity. Having gained the center
in comparative safety, we made a long stride, and
leaped on what seemed a portion of the floating,
jointless bridge ; when, horror of horrors, in a moment, a flash of time, yet long enough to awaken us
to the imminent danger, a sudden, writhing plunge,
and fierce splash of the living tail of the supposed
log, and we went swirling into the water among
the slabs, from which our fright and frenzy speedily
brought us, somewhat wet and bruised, but exceedingly thankful. The deceitful " old log " was
a few yards away, slowly sailing toward a circle
of weeds and tall grass, with his round eyes just

above the water's surface, and evidently surveying
us, while we shook our impotent fists in rage at his
head. This clump of grass and reeds, we concluded, must have been a regular alligators' nest;
for, besides this big fellow that had so incontinently
upset us, we counted three others adjacent, and all
within five rods from the spot on which we now
stood. That alligators are comparatively harmless
to people, we had been told, and had now some
reason to believe: still, we should not like to encounter them in the water. A hundred years ago,
they were much more plentiful than now along this
river. Bartram, the younger, relates having seen
them everywhere along the river and the banks
of the St. John's. He also mentions their trying
to get into his boat when it was tied to the bank;
but he did not suppose it was their intention to
destroy it, but merely to possess its contents.
They will give fearful battle among themselves
frequently. The two master animals of different
communities have, like champions, been seen by
Bartram to engage each other in a terrible contest;
the others being silent spectators to the close of the
scene, when the friends of the winning combatant
would roar with a victorious clamor. We have

never heard of any person being destroyed by them,
or injured except when imprudently approaching
them on shore while wounded. We might have
captured one had Mr. Murray's intrepid guide, of
Adirondack celebrity, been along at the time of
crossing the floating bridge, we should have had no
difficulty in securing the alligator for preservation
in some historical or antiquarian collection ; since,
if he had had the same presence of mind as when
he caught and held Murray's deer, the question
of capture could not have been a doubtful one,
as his opportunities to *hold on* would have been
greater! We should at least have surrendered
with as much *nonchalance* as Mr. Murray, our
claim to the beast.

We returned to the village with courage una-
bated, but our bump of *caution* had been by the
morning's adventures considerably enlarged. Procuring a team of a villager, who kindly consented
to accompany us, we proceeded, but on quite
another road, across the pine woods, dodging the
obstructions, and, notwithstanding, making good
time.

CHAPTER VII.

CENTRAL FLORIDA.

PICOLATA.

To Picolata from Green Cove is but a couple of hours' sail; fare, one dollar. The only impression made by this short trip on the river was its lake-like appearance. Picolata figures conspicuously on most of the maps as an important town; but the reality is so absolutely nothing, that it is truly laughable to think that it should ever have had any name beyond, perhaps, " Brown's Farm," or " Smith's Stables." An inventory is as quickly given as any thing: one beer-shop, one comfortable farm-dwelling, and a few rude stalls constructed and occupied for the use of the stage-lines to St. Augustine is all. Beyond being the nearest point on the river to that old city, and the place

105

where passengers destined for St. Augustine disembark, it is of no importance. We doubt whether
a *respectable* pauper at the North would take the
whole thing as a gift, and be obliged to stay upon it.

On our return trip, we shall get off here, and go
by stage 18 miles through the pine woods to the
old city. But, for the present, we propose to take
the reader to the Upper St. John's, the gem of all
rivers.

Making no note of the several small landings,
where our steamer, " The Darlington," called for
wood and delivered the mails, we are only interested in the broad river which moves toward
the sea, and watch the flocks of wild ducks as they
fly athwart our bows, and dream of the fascination
which this land had for those who were its first and
rightful owners. Here were fish, flesh, and fowl at
their feet, in quantities sufficient to supply their
demands; and no wonder they disputed the inva-
der's right to the last acre with an unequaled
desperation and heroism.

The sky, which has all day been bright and
sunny, is fast becoming overcast with threatening
clouds, that seem to gather rapidly in the south-
west: an hour later, and the slanting rain warns us

off the deck, and we seek the cabin of our good ship "Darlington." Perhaps a word as to the boat and its officers will be welcome, since this floating hotel is to be our home for a day or two. First, as to the boat, which has accommodations, we should think, for not over forty passengers, and this trip she was not crowded; of fair speed, and built not unlike the Western river-boats, — giving up all of the first deck to the machinery and freight, while the upper deck is devoted to passengers; having two saloons, one quite small for ladies and children, while the main saloon is frequented by all; around this are ranged the state-rooms, while at meal-time it serves as a dining-hall. Its chief officers are Capt. Jacob Brock and Admiral "Rose." The first sails the boat, assisted by a pilot, while the *Admiral* is in actual command. Since she (the Admiral) is better known along the river and by travelers by the name of "Rose," we may follow the custom, though unquestionably the title is due her. She was not born to bloom and blush unseen, — not she. Her every word of command might be heard ringing out sharp and clear above the noise and confusion at every landing. Her word was law : her orders were instantly executed by every officer below the captain ; and with him she held occasional conference, in which it could well be observed, by the tone and gesture, that it was she who gave the final word. A stout-built athlete of two hundred pounds, of medium height, a full piercing eye, regular features, and with a peculiarly commanding voice, and look of one whom experience had taught that life was a battle, and they who made the best fight won ; a flow of ideas, logical and pointed, with a half ubiquity of presence and an equal mixture of the Seminole and Anglo-Saxon wine of life, and you have before you the portrait of "Admiral Rose." She was a study and an object of no little interest to all.

Our captain was formerly from Connecticut, we believe : he at least married there, but has lived long enough away, buffeting the world, as at times to quite forget the pious teachings of his early youth. However, he is a very efficient, and, so far as we observed, a very obliging officer. His native energy and tact have served him to excellent purpose, we judge ; since he is now not only the master but owner of his boat, as also of the hotel at the end of the route, and other property in the State of considerable value. He informed us that he commenced his steam-boating career on this river nearly fifteen years ago, and that this same " Darlington " had been in service most of that time, and she still seemed good for half a score of additional years.

PALATKA

is reached late in the afternoon. This town is on the west side of the river, and is larger than Green Cove: indeed, this is the largest place on the river above Jacksonville, and is about one hundred miles south of the river's mouth. Its population can not be over half a thousand. Occupying a high and dry site above the river, extending a quarter of a mile along its banks, it appears to be a pleasant, thriving village, with eight stores, including a fine drug-store, two large hotels, two churches somewhat neglected, two steam saw-mills, and several fine residences, with the balance plain and scattering. The stores seemed well filled with a large stock of goods, from which an extensive back country had to be supplied. Palatka is a kind of steamboat head-quarters for the Upper St. John's and its tributaries, and is the terminus of the Savannah and Charleston lines; namely, the "Dictator" and "City Point," running from Charleston to Jacksonville, thence to this place. This line is known as the "outside line," since it takes to the ocean between the mouth of the river and Savannah. The other line, composed of the "Nick King," and "Lizzie Baker," runs along the inside passage, off the sea-coast and the mainland, to the mouth of the river, thence to this place, and is known as the "inside line." The two lines make four round trips each week. Then there are the river-lines, which run from Jacksonville up to this place, and beyond, to the head of navigation. These, with the lines to Dunn's Lake and Ocklawaha, which run weekly, make daily arrivals at this point. We counted four steamers tied up to the wharf the evening of our arrival. Most of them lie over here all night, passengers having the privilege of remaining on board or going to that excellent hotel, The Putnam House, kept by O. E. Austin, of "Down-East" enterprise.

Palatka has a newspaper, "The Eastern Herald," a *weekly*, of circumscribed dimension and circula-tion. It gave us a surprise to find such an article here, and helped to drive away an impression which some how had taken hold of us, that we were in the woods.

Many of the recent settlers or new-comers are from Auburn, and the vicinity of Central New York, — persons whose health has been more or less precarious, and who have found benefit in a residence here. Though their speculations, we understand, have not proved as remunerative as was anticipated, still, some have taken hold with a resolution that doubtless will, in a few years, "prove some things as well as others."

This place affords a delightful climate most of the year, and is in our opinion as far south as invalids need to go, though many do go farther, and perhaps with advantage.

Early vegetables can be had here in March; and, indeed, we have eaten here, new peas and potatoes this month. What a strange experience for the Northerner, who has come away from the frozen fields and gardens of the North, where these things do not ripen, on the average, before the 1st of July, over four months later!

Opposite this place, on a high bank, stands a beautiful orange-grove, with a few oranges left from the December frosts. Not since 1835 has

there been a frost to kill the trees or destroy the
crop till last December (1868), when a heavy
frost destroyed most of the fruit, except in the
southern half of the State; but the trees themselves received no damage. This grove opposite
Palatka contains five hundred trees, and is the
largest we have seen in the State. They had
commenced to gather the fruit before the frost;
and a single tree had yielded sufficient to net
sixty-five dollars, while some trees had over a
hundred dollars' worth of oranges.* This grove is
neatly kept and very beautiful, and forms the
chief attraction about Palatka.

The character of the river above Palatka is in
striking contrast with that of the portion below.
The town stands nearly at the central point of its
length; and that portion below, to its mouth, may
be rightly designated the " Lower," and that above
the " Upper " St. John's; though, strictly, it is
some miles above this town where the river con-
tracts rapidly its banks, and discloses those peculiar
features which make this peerless river of the
South the admiration of every lover of the pictur-
esque and beautiful. The water is clear, and the

---

* Letter of Frank L. Burr, Esq., in Hartford Times. Our thanks are
due to Mr. Burr for copies of his letters.

banks present a richer and more luxuriant growth
of semi-tropical vegetation; and then there is no
longer that uniform and monotonous landscape, as
below, while each bend of the river presents new
beauties in delightful and endless variety. Seven
miles above Palatka, and we come to the outlet of
Dunn's Lake, lying a few miles to the south-east,
which is reached by a small steamer once a week
from Palatka. The lands around this lake are highly esteemed, being favorable to the cultivation of
fruit and vegetables. Several Northern gentlemen
own interests there. Our friend David Clark of
Hartford is among these. He is a half-owner in
a wild-orange grove of several hundred acres,
which is now being transplanted into rows, and
budded with the sweet fruit. In a few years, he
will probably possess one of the finest groves, as
well as one of the largest, in the State. The
entrance to Dunn's Lake and the mouth of the
Ocklawaha River on the St. John's are separated
by a distance of about ten miles. The Ocklawaha
is a navigable stream for some one hundred miles,
running in a tortuous manner westward and
southward, passing through occasional lakes, with
its source in the everglades of Southern Florida.

It is a very narrow but a very deep stream, and is over two hundred miles in length. The region of country through which it runs is thought to be finer in many respects than that along the St. John's. Some distance up the river is the famed Silver Spring, a great basin of surprisingly clear and deep water, around which it is supposed clustered a populous village of the aborigines numbering over six thousand inhabitants. All this vast region was once thickly settled by these people. We regretted that it semed impossible for us to explore the country bordering on this stream; but that would have consumed more time than it was possible to give.

Near the mouth of the Ocklawaha is the old town of Welaka, once the site of an Indian village, and afterwards a flourishing Spanish settlement: now scarce a vestige of either of these populations survives to tell of their existence, so completely has time leveled every thing. Volusia, farther up, and beyond Lake George, is another city of the past with an unwritten history, though known to have been an important point when both the Spanish and English held it. It was on the line of travel, and one of the principal points for crossing from St. Augustine and the Musquito Inlet country, on the east coast into Middle and Western Florida. It occupies a high site; and the opposite shore stands boldly out, — quite in opposition to the theory of one writer, who asserts there are no points opposite in this river where both sides are high and dry. Indigo was extensively cultivated in the surrounding country a century ago, while rice and cotton received considerable attention.

Before reaching this section, however, we pass through a large lake, some sixteen miles long and more than half that distance across, known as Lake George. On entering it, several islands are seen, and one of considerable size, called Rembrandt's, that contains near two thousand acres, on which an orange-orchard stands, and a two-story frame-house, which is rather a novelty, since few houses of any sort are seen on the river, so much of a wilderness is it. Even the shore of this lake is scarce broken in its entire circumference by any settlement.

The St. John's is separated from the ocean its entire length by a comparatively narrow strip of land of not over forty miles in average width. Its banks are mostly low, very apt to be swampy, and are densely covered with the primeval forest, — live-oaks, water-oaks, bay, gum, and magnolia, hung with the long waving drapery of solemn moss, while the beautiful palm with its crested crown sentinels its shores. Beneath and interwoven among these are the wild jessamine and creeping vines, lacing the undergrowth over with an unbroken verdure, and rising from the water to the tree-tops. Then a dell appears, festooned with climbing and o'er-arching growths, gracefully curtaining its entrance, and tempting its beholder within the half-hidden sylvan retreat; farther on, and a broad expanse, or savanna, is seen, bounded by the distant forest, with here and there a solitary palmetto standing alone like a plume. Flowing for some distance between banks of living green, the river suddenly widens, and we come to the placid waters of the lovely Lake George, across which we sail to the continuing channel beyond. By some strange instinct and guidance, as it were, we find, hidden behind the tall grass and floating islands of this really fairy land, the sought-for channel, though a half-score of equally promising openings had nearly tempted us astray.

Again and again, this bewitching transfer is

made from lagoon to river and from river to lagoon and lake again. The shores here and there close to within a hundred feet, and is constantly turning the sharpest of angles, running back beside itself, fascinated with its own beauty as it were, then whisking suddenly about, as if on a wager to outdo itself. On either hand, glimpses are had of bays and mirrored waters, whose surface and shores seem likely never to have waked to other echoes than those of the wild birds that inhabit them. Again the river seems lost in a bewildering perspective of silver streamlets, separated only by some narrow knife-blade of meadow-land and flowers; but up them they are all verdure-banked and moss-hung and vine-curtained and flower-bedecked, — the very picture of tropical profusion and summer loveliness. The swelling waves from the passing steamer lift and move away islands that just now were, but which a moment after are seen drifting with the current and breeze in fragmentary sections, each a living, floating, tropical bouquet.

On these shores were the beautiful homes of those brave red men who were so cruelly hunted by the bloodhounds of Jackson, the story of which is a dark and cruel one, and causes marvel at that inscrutable Providence which permits the inhumanity of man to his fellows.*

The weather is delightful; and with this odor of summer in the air, the spirit of Beauty everywhere around us, in the voiceful forest and bright waters, we are soon lost in silent contemplation.

Along the banks, and at the more elevated portions, crops out just under the soil a crust of a foot in thickness, of snail-shell formation; and also the coquina, a species of shell-rock, was seen. Wherever this shell is, the soil is fertile, and especially favorable to the growth of the orange, which may be found in a wild state on the banks and lands adjacent.

The birds and wild fowl seem to have possession of this land, with no one to question their right or to make them afraid — of gunpowder.

One half of the country upon the Upper St. John's is but a wide, extended plain, half soaked

---

* The Administration, it is reported, sanctioned the use of blood-hounds in driving the Indians out of Florida.

THE HOME OF THE HERON.

in water, and covered with the tall reed-grass, trampled and broken down by the gigantic waterfowl which inhabit them. Great flocks of herons rise and circle the air, flying in long lines, and making at early morning the sky vocal with their wild song; then descending in regular column, light either upon the plains to feed, or on the tops of the tall cypress, venerable with years, and bearded all over with the long, waving gray moss which here drapes the forest everywhere. Then again the white curlews, so handsome and so shy, look like troops of embodied spirits, — they are so pure and bright in color. The tall cranes, standing singly here and there, — to birds what the palm is among trees, — start up with a dull and heavy stroke, beating the rank growths heavily for rods along. We have seen them, so large, and so suddenly startled, causing them to fly forestward, where their sweep was so pent up as to prove an ignominious failure, coming down with a piteous cry or squawk, as if for mercy. The pelican is not so often seen. This bird is among the largest in the State, and chiefly frequents the low savannas, but seldom in pairs. Loons of various species inhabit these waters: one is of a somewhat remarkable character in color and habits, possessing a sharp bill, and a long slender neck of dun or cream-color, and the body and tail of a jet black, with the latter white-tipped. They seem fond of airing themselves on limbs just above the mirrored waters, whence they doubtless contemplate their figures as reflected by the polished surface beneath with as much interest as any modern belle adjusting her toilet in her boudoir. If discovered, they suddenly drop into the water and quite disappear, and in a few moments may be descried rods away, skimming the surface, with nothing but their long necks and the tips of their tails visible above the water. They present the appearance of a water-snake at this time, and are therefore known among the people as snake-birds. In paroquets, the colored people find an article of trade and commerce. Numbers of them were brought on board at one of the landings, and were speedily snapped up by the *pater familias* of admiring and persuasive young ladies, who thought them "*so splendid*," that a couple of dollars a head was readily paid for them, when, we believe, in the New-York bird-markets they can be had for half that sum.

But the poor birds were short-lived. A subsequent investment in a young alligator, of dimensions suitable for a cigar-box, was *so much more* " *charming*," that the brilliant but woe-begone looking paroquets died from exhausted admiration! Eagles are frequently seen, as also great fish-hawks that are often mistaken for the former. This land is to them one of unbounded resources, and they wax fat, and lord it with keenest relish.

Innumerable smaller birds infest the woods, and chant their grand oratorios at dawn, while their inexplicable chatter fills up the day. But the queens and glory of this tropical land, in bird-music, are the somber-robed, bright-faced mocking-birds, who for a moment halt in their aimless ways, scattering glittering and delicious gems of song through all the woods. What silence ensues in the vast company of choristers and imitators when these Jenny Linds appear, eloquent with a melody all their own! It seems, while we listen, as if this were " Paradise Regained," such are the sights and sensations as we sail up and down this upper river, so tropical and so unique. It is one long dream of beauty, with choicest pictures of Oriental luxury and repose. Here let some unknown artist come, with the hand of genius to imitate these scenes, and his work shall ever rank with the most favored, as it would be among the most attractive in the land.

CHAPTER VIII.

THE UPPER ST. JOHN'S.

LAKE MONROE.

OUR last chapter brought us to Lake Monroe,
which stands at the head of regular steamboat
navigation on the St. John's River.   Entering this
handsome sheet of water, of about five miles in
diameter and nearly circular, we are able in clear
weather to at once discern its two principal points
of interest.   The first is Mellonville, on the western shore, formerly the site of Fort Mellon, built
during the campaigns against the Indians.   This
point was originally, however, a trading-post with
the natives, and quantities of furs were annually
brought here.   The town itself is nothing worth

123

naming, beyond a wharf, small warehouse, and an humble dwelling combined. It looks desolate ; and yet we see a full half-score of settlers from the near interior, in waiting on the pier to obtain such articles as they had severally ordered. We were detained here a half-hour in discharging freight, consisting of flour, meal, some little machinery, implements of agriculture, and other minor packages of store articles brought from be- low. The country adjacent is little cultivated ; but back some twenty miles and more it is of a much better character, and has a prosperous settle- ment of quite intelligent and industrious planters. The soil here seems sandy, as it appears every- where away from the hummock-lands. The mails of " Uncle Samuel " are exchanged ; and the steamer, veering round, steers straight across to the eastern shore, and to the town of

ENTERPRISE.

A half-hour of steady work of the engines brings us to its wharf, extending ten rods into the lake. Our eyes surveyed the place a score of times ere the gang-plank could be made ready ; and we were near having convulsions at the thought of what we had pictured it and the reality before us.

A half-dozen buildings is the sum total of Enterprise, about which so much is heard by travelers on the river. A feeling of disappointment is inevitable ; notwithstanding the traveler soon accommodates himself to the fact of utter barrenness in a country which has been for a century forgotten ; yet, that at the head of navigation, the termini of steamships and travel, there should be found but *one house*, is beyond all belief! This is, however, the case ; though a new dwelling was in course of erection during our visit. The hotel, the Brock House, the center and circumference of this *city*, is also a disappointment; but it is in this case an *agreeable* one. From Jacksonville up to Enterprise, no hotel equals this. It is a genuine Northern-looking hotel, such as you may see at a watering-place on the seaboard, like, for instance, the old " Watch Hill House " at Watch Hill, in Rhode Island. It stands broadside to the lake, one hundred and ten feet long, and two stories and a half in heighth, with a veranda its entire front, broad and airy. The house is well painted, and has attractive green blinds, and comfortable accommodations for upward of fifty guests. It is also well kept, by a gentleman from Maine, who makes himself a polite and pleasant host. The table during our stay was well supplied with meats and fish. The night of our arrival the landlord set before us beef-steak, ham and eggs, roast venison, and fish. The venison had been caught that morning; and before sunset, within a mile of the hotel, another buck was shot, a fine large one, off which a large party breakfasted the next morning. The woods are full of game, and the lake swarms with a variety of good table-fish. Trout, perch, and mullet are found in nearly all the lakes, while bass, sheep's-head, and bream abound the entire length of the river. Shad are taken in their season, as in other rivers to which they have access. Probably no river on the globe of equal size affords a greater supply of food for man than this. The most of our rivers are so dragged and hunted that fish are scarce; and it would seem only in fancy that we could, within a four days' journey from the "Empire City" of New York, find a river full of fish, and as plentiful as though its discovery was but yesterday, instead of its being probably the first river entered on the continent by any European.

Why, you may see them fairly dancing on the surface of some of these lakes. Bartram, a hundred years ago, tells a story, that, at Battle-Pass, he witnessed a sight which those who know this river can readily credit. He was working his way up stream, when, observing the bloody appearance of the water, he hastened onward, and soon came to a narrow pass, where he was wonder-stricken at the sight that met his view. A vast swarm of fish was crowding by this point: they moved rapidly, yet the river was gorged and fairly dammed. On either shore, and in the stream, were a score of huge alligators, with distended jaws, crushing and destroying the helpless creatures with an astonishing rapidity.

The fish caught in the Lower St. John's will average from one to forty pounds; while in the Upper River, from a half pound to fifteen is about the average weight.

Of wild fowl suitable for the table, ducks are the most abundant. There are several varieties. Almost the first thing seen in coming into the river, at its mouth, are flocks of ducks; and they may be found all over the State, but especially on this river, which seems their favorite home. Wild geese are common in the season for them, but yet they go farther south as a rule. Indian-River country, some thirty miles eastward, is more frequented by them. Quails and partridges are often seen; but the hawks and other birds of prey, so numerous here, keep them from increasing their numbers to any great extent.

Enterprise is the paradise for sporting-men. For invalids to discuss the respective merits of this or that place is proper, but there is no question where the huntsman or sporting-man should go. This place is their true head-quarters: none other equals or compares for a moment with it. Once here, they have a central point from which they can move at leisure, and return for repairs and rest. If it be fishing, the lake is before them, and they need not angle around half the day with a " fly " and no fish. It may be all very well for those who like it, to stand braced against a tree or sit silent as death in a boat, perspiring at a nibble, and catching nothing but a cold; or possibly, if very lucky, bringing home a half-pound trout or two, scarce enough to smell of, much less to satisfy a ravenous appetite. This is the usual modicum of luck and result elsewhere, the Adirondacks not excepted. But, when *we* go fishing, we like to do a good business, both for ourselves and friends; and here it can be done, for instead of a bag it is your boat full. The glory and delight which thrill the nerves of such fishermen is royal and worth experiencing.

Lakes Jessup and Harney, above, are also well stocked with fish of excellent quality, which are easily caught by nets, hooks, and spears.* A very attractive sight at Enterprise is the orange-orchard standing at the left of the wharf, and owned by Capt. Brock, who is also proprietor of the hotel, and, for that matter, of every thing else of any account at Enterprise, not excepting the steamboat line which brings you and on which you have to return. This grove is a very handsome one, and covers a couple of acres across the roadway from the hotel. The trees are of fine size, and very pretty in appearance. There was some fruit still hanging, large and golden, but valueless, its juice having dried away, though the exterior was fair and plump-appearing. In the tops of the trees, blossoms could be seen at the time of our visit, and another fall they will again be loaded. There

* Letter of Solon Robinson, in Tribune.

9

are a few lemons in this grove, and also some of
the sour orange. These doubtless are left for
variety's sake; though the lemons are as profitable
and useful as the orange, but not so commonly
grown.

Capt. Brock built and founded this new
"Enterprise," and ran his line of steamers,
determined to make it pay; and he is now able to
realize the fruit of his persevering toil. The
past year the house has been crowded, the appli-
cants being double the number that could be
entertained. Old Enterprise is, as we have already remarked, about a mile above, on the lake-
shore. This was formerly *the* place; but Brock,
having the steamers, had the power to establish
a successful rival. Old Enterprise is nevertheless
the place where Brock's hotel and orange-grove
should be, as it is higher, and has from fifty to one
hundred acres of cleared lands in a condition for
cultivation; whereas at Brock's Enterprise, beyond
his garden and orange-orchard, there is not one
acre of thoroughly cleared land that we observed.
Old Enterprise has an orange-grove too; but it is
scattering and in no wise particularly attractive.
Dr. Stark, its occupant and owner, though a

Southern sympathizer during the war, as we learned, is, nevertheless, a very courteous and intelligent gentleman. If a hotel were erected on this old site, with bathing-rooms supplied with water from the large spring just back of his grounds, — of which we shall hereafter speak, — it could not fail to take its share of the winter travel.

Enterprise is the point at which passengers land who are bound for New Smyrna, Musquito Inlet, Hillsborough, or Indian River on the sea-coast, and directly eastward, a score and a half of miles from this place.

There the banana, fig, and pine-apple are seen, and the coffee-plant grows wild and luxuriant. You may revel in the delights of earth, and wish, as De Leon did three hundred years before you, for some fountain or fruit of which to taste, and live for ever in the possession of youth, giving beauty a second consideration entirely.

The climate at Enterprise is perceptibly milder than at any point below. This is not greatly to be wondered at, perhaps, since every hundred miles south, when so near the tropics, makes a distinguishable difference, — far more marked than the same distance southward in the Northern States. The winter here is not unlike in temperature the month of May in New England. We find that fires are comfortable and necessary evenings in February at Enterprise, and they are also equally necessary in June in the Connecticut Valley. The health of people all over our country would be much improved, and many cases of fevers prevented, were small fires started in the evening, during most of the spring and fall months.

CHAPTER IX.

CELEBRATED SPRINGS.

THE springs throughout Florida are numerous, and many are quite remarkable. They form one of the wonders of the State.

At Enterprise, every visitor seeks the famous spring of this locality. It is nearly a mile southward from the hotel, and we have the choice of two ways by which it may be reached; and they are quite like the Irishman's roads, for, whichever one you take, you will be likely to wish you had taken the other. One is by the beach, where little swails and sluices, emptying into the lake, have to be leaped in wet weather; the other is back of the hotel, — a winding cart-path among the low scrub palmettoes and tall long-leaved pines, then through a clump of magnolias, gums, water-

133

oaks, and sweet bays, with the towering palm shooting straight into the air as an arrow for ninety feet, capped by its globe of green. We pass a wilderness of jungle and forest and vines and wild flowers, without name and number. The air is soft and tropical; and the scene is what we might easily fancy a Brazilian forest to be, lacking only in the flashing eye-balls of some crouching animal, to light up the jungles by which our circuitous pathway leads. Turning suddenly aside, across a clearing into the woods beyond, and we stand at the spring. It is circular in form, and nearly eighty feet in diameter; and its surface is as smooth and unrippling as if congealed. The water is of a delicate green and quite transparent. Its depth is said to be full an hundred feet. No living thing was seen in its waters, which are sulphurous, though not markedly so. Its taste, however, was not especially pleasant; and we should prefer good clear spring-water to it, for purposes of health. It might serve a good use for bathing; but it is, in all respects, far inferior to the healthful crystal spring at Green Cove, which is employed by citizens and strangers for every domestic use, and as a healthful tonic. This

spring is not now used, and we observed no signs
of its having been by any one. It is simply a
curiosity, and is visited as such. An expert
swimmer might bathe in it, but none others, since
its shores are as deep as its center, and it is quite
forbidding in view of its depth and color. A
small outlet, six feet wide and as many inches in
depth, carries the wastage waters to the lake, a
third of a mile distant.

Our party grouped themselves for a few
moments, long enough to hang this Oriental
scene in memory ; then, gathering a few wild
branches as mementoes to our friends away on the
banks of the White Water and the Ohio, the
Hudson and the Delaware Rivers, returned, *via*
Old Enterprise, to our hotel.

We have spoken in a preceding chapter of the
beautiful Silver Spring on the Ocklawaha River.
That is a great basin of surpassingly clear and
deep water. Springs of salt water, and springs
of mineral water, are not uncommon in various
sections of the State. They are all usually of a
tepid character, standing at about seventy-seven
degrees, we should judge, having frequently bathed
in them, although we never subjected any of them

to the test of the thermometer. These springs are
usually near the rivers or lakes, and occasionally
they have been discovered boiling up in the lake-
waters. Mr. Solon Robinson relates having passed
over one while sailing in a small boat along the
shores of Lake George, and gives currency to the
report of springs being found off the coast of St.
Augustine, which is not unlikely, when we con-
sider the peculiar character of the peninsular
formation.

Right in the middle of the St. John's River
up   toward Lake Harney, there boils up a
tremendous spring, which makes a conspicuous
turbulence on its bosom, — enough to shake the
steamer; and, in ordinary stages of the river, this
spring lifts its waters visibly above the surface.
The captain stated he had sounded it to a depth
of three hundred feet, and found no bottom!
Lake Jessup, on its western shore, has several
large sulphur springs. This lake can not be
entered by a boat drawing over three feet of
water. It is some seventeen miles long by five
in width, with shallow waters generally. Its
shores are low and marshy, as a rule; but here
and there are dry shell-banks on which the wild

orange grows. Some miles to the west, however, the land is better, being higher and
healthier.

One of the most famous springs in the State,
perhaps, is but a few miles south of Enterprise. It
is known as the Blue Spring; but among the inhabitants, here, all springs are called blue springs, little
mattering what the real shade of color may hap-
pen to be. This spring is on the east side of the
river, near Lake Berresford, and empties into the
St. John's a half-mile from the steamboat landing.
It is perhaps the largest spring in the State: the
quantity of water which issues from it in an hour is
enormous. It forms a river of itself, one hundred
and fifty feet wide and six deep; sufficiently large
to admit the passage of a considerable craft. The
water boils up out of the earth as though from a
boiling caldron of four-score feet across. An
excursion party from Jacksonville tried to row a
boat into the center of this boiling kettle, in order
to take soundings, but were foiled, after several
earnest efforts, in consequence of the violent motion of the elevated surface.

A trip to this spring is a pleasant and easy excursion with the little steamer " Hattie."

With a quaint account by Bartram of a spring he observed, we shall close what we have to say under this head.

He says, . . . "On my right, and all around behind me, was a fruitful orange-grove, with palms and magnolias interspersed; in front, near my feet, was the enchanting and amazing crystal fountain, which incessantly threw up, from dark rocky caverns below, tons of water every minute, forming a basin capacious enough for large shallops to ride in, and a creek of four or five feet depth of water, and nearly twenty yards over, which meanders six miles through green meadows, pouring its limpid waters into the great Lake George, where they seem to remain pure and unmixed. About twenty yards from the upper edge of the basin, and directly opposite to the mouth, or outlet, of the creek, is a continual and amazing ebullition, where the waters are thrown up in such abundance and great force, as to jet and swell up two or three feet above the common surface; while sand and small particles of shells are thrown up with the waters near to the top, subside with the expanding flood, and gently sink again, forming a large rim, or funnel, round about the aperture, or mouth, of

the fountain, which is a vast perforation through a
bed of rocks, the ragged points of which are
projected out on every side. . . . The waters
are so extremely clear as to be absolutely diaph-
anous as the ether; the margin of the creek (?) is
shaded by a great variety of fruitful and flowering
shrubs and trees, the pendent golden orange dancing on the surface of the waters, and the songs of
merry birds vibrating through all the trees. Below,
innumerable bands of fish are seen, some clothed
in the most brilliant colors; the voracious crocodile,
stretched along at full length like the trunk of a
tree; the devouring gar-fish, trout, all the varieties of gilded bream, catfish, sting-ray, skate, and
flounder, spotted bass (trout), sheep's-head, and
drum, all in separate bands, and circling in peaceful evolutions with none of their usual signs of
enmity. . . . See whole companies of these fish
descend into the abyss; they entirely disappear:
are they gone for ever? . . . . Looking intently
to observe if they returned, when, lo! they were
seen emerging from the depths, apparently at a
vast distance, as they at first seemed no bigger
than flies or minnows; now gradually enlarging,
and rising rapidly, they ride forth with the elastic,

expanding column of crystalline waters, and gently
move each to their kindred tribes to re-form and
renew the sport.   This scene seems unreal, since
it appears that you are at times able to almost
grasp the fish as they rise toward the surface,
when in fact they are thirty feet away."

Though we have spoken in a previous chapter
of the spring at Green Cove, yet to omit mention
of it here would be an injustice, not for any thing
remarkable about it as seen by the beholder, but
for its high value as a remedial agent.   In this
particular, it excels all others with which we are
acquainted.   Its water is pure and pleasant, and
has the credit of having cured some invalids and
greatly benefited many others.   Its depth is such
that any one may bathe in it without fear, and the
use of the water increases the appetite and
strengthens the system.   Indeed, this is the only
spring with which we are familiar in this State
that has any reputation as a curative agent.
Being located within the village of Green Cove on
the St. John's River, it is accessible to those in
feeble health as well as the robust, and all may be
benefited by its use; though, in this matter, it is
proper for the invalid to consult with his medical

adviser, since in some cases, doubtless, its use would be injudicious.

Springs abound in all portions of the State, in the western as well as the eastern section; and they are all of more or less interest as curiosities, and will well repay the tourist.

CHAPTER X.

ALLIGATOR-SHOOTING ON THE UPPER ST. JOHN'S.

The Steamer " Hattie."—Sportsmen on Lake Harney and beyond. — What they bagged. — The Grandfather of all the Alligators. — Difficulty of killing them. — Their Display of Gymnastics. — The Party camp for the Night. — A Thrilling Scene. — Our "John." — A Wild Night of it. — The Morning Alligator *Reveille*. — Reflections.

THERE is a small steamer, the little "Hattie," which runs to accommodate hunting or pleasure parties above Enterprise, and can make the run to Lake Harney and back in a single day. Toodles would have set an inestimable value on this pocket-craft, as, indeed, do scores of people who come to Lake Monroe. She can be chartered for any length of time ; and, if there are a half-dozen to go, it is not expensive, as the cost does not exceed the price of board at a first-class hotel. We know of a hunting-party composed of New Yorkers and Baltimoreans, who proceeded to Lake Harney, and beyond, as far as it was possible with safety to take the little steamer. All were well equipped with arms, ammunition, and provisions. Their brightest expectations were fully realized. Game of all kinds was found in great abundance. They shot twenty-seven deer and one hundred and sixty-two alligators, not to mention any thing of smaller game. They returned bronzed and delighted, and with an increase of avoirdupoise. All who come as high as Enterprise should come to take a hunt, and try the lake-fishing.

Alligator shooting is among the finest of sports, and besides is strange and pleasing to most visitors, who, fresh from Northern snows and ice, take to the fun with a rare relish. A bright day is necessary: such was the one chosen, when, with a small company, we set sail for the mouth of the river on the south side of Lake Monroe. "The Hattie" puffed vigorously across the lake, but, entering the river, "slowed," and quit her "wheezing:" all assembled on deck, and, with rifles plenty, we had not long to watch for the amphibious monsters. They are soon seen swimming across our bows, or lying lazily on the shores as is their habit; when, crack, bang, and pop, go the rifles, and an alligator flounders down from the shore into the water. Some times a half-dozen bullets would strike the monster at about one and the same time; and the huge beast would writhe in agony, lashing the reed-grass with his powerful tail, and lurch into the dark waters with a despairing and desperate plunge. Again a lucky or scientific shot back of the fore-arm would bring the scarlet tide, when with a sudden whirl upon his back, and with uplifted, quivering limbs, he was off to the land of shades without further ado. These cases were valued, since it afforded an opportunity to draw up to the shore, land, and decapitate the giant, bringing his head away, not only as a trophy, but for the purpose of practicing at our leisure a little dentistry on his teeth; and in time these became, in the hands of some expert, a beautiful whistle, or, carved into various forms, were prized as charms.

At first our shots were wild, owing to the excitement: soon all became practiced, and could aim the deadly missiles with rare exactness; yet comparatively few were killed outright, though numbers were wounded. This was attributable to our finding the majority of them in the water; for, hearing us, they would slip off of the bank and float on the surface, when nothing could be seen but their head just above the water. It takes a keen and practiced eye to detect an alligator, they so closely resemble a rotten log half submerged and motionless.

If a ball enters their eye squarely, they are finished; but to do this is difficult. A good loud rap on their heads can be given them, however; and the way the water flies when they are thus hit and hurt is both amusing and wonderful. It is beaten into a perfect foam ere they plunge from sight. One was a mighty fellow. He was seen at a distance, lying on the marshy banks at the bend of a river, and appeared the great grandfather of all the alligators. Experience had taught him it wasn't safe to wait for excursion parties, and he began to move " early; " but our crack marksman drew a bead on him, and at the vital spot. Halting to take a final observation before making a plunge, the sharp crack of the rifle, and a dull thud for the echo, told the story. His delay had proved fatal, and he went down to his grave with a gorgeous display of gymnastics. Suddenly throwing up his fore-arm, he rung down the curtain and his own life at the same time.

10

The day passed rapidly by, and the party decided to tie up for the night, and continue the sport another day. Passing on to a dry shore, we made our camp in a grove of the wild orange, and under the towering palm. It was a lovely place for our purpose. Those who preferred remained on board, while full half of the company pitched the fly-tent, and prepared a faggot fire, wherewith to cook our fish-supper. This was soon accomplished, and as soon disposed of; for what will sharpen appetites better than a day upon the water, full of excitement, frolic, and fun! A few sauntered back inland among the maples and magnolias for game, and were not long in starting a flock of wild turkeys from among the tall trees, where they were about going to roost. A little management was required to secure a good shot at the cock-turkey; two of us agreeing to fire at a given signal from the third, and he to follow with his rifle if we missed. The distance was considerable, and at quite an elevation; but, at the word "Fire," the bouncing gobbler came to the ground, the balance of the flock scaling away from our sight and reach. Shouldering our Thanksgiving-bird, we reached camp, and

were congratulated heartily by our companions.
Ere long the moon, full and clear, swung itself above
the tree-tops, and all joined in some stirring
melodies that woke the echoes of the woods and
surrounding shores. A few branches and leaves
from the palmetto served, with a blanket over-
spread, for our bed.

In the night we were all aroused by the hoot-
ing owls; and such a din! A dozen barking pups
would have been out-matched. "John" decided to
give them a salute with his double-barrelled gun,
and, rising to reach it, imagine his feelings, and
ours as well, for it was but a moment after his
discovery that he *telegraphed*, by a vigorous kick,
the news to us. Not over ten feet away was a
good-sized and every way respectable-looking
"gaitor," paying his addresses to our proposed
breakfast, the turkey, which was hanging to a
low limb of an orange-tree close by. How long
he had been there we did not know; or how long
it would have been ere he would have snapped up
our poor gobbler, or swallowed a pair of boots
with John's feet in them! it would be difficult to
determine; but that crocodile was hardly ex-
pecting such an emphatic greeting. Our rising

had evidently checked his progress toward the
turkey; for now his eyes were turned fully upon
us, as if unable to make out whether we were
friend or foe.   His suspense was brief: a moment
fully undeceived him.   Such a crash of minnie-
balls and buck-shot as landed in among "his
horns" was totally unexpected.   His manner
indicated as much; for in a trice he performed the
" Grecian," and tilted endwise into the river, with
such a roar and splash of water as we never
heard before.   The owls were silenced, and dis-
turbed no more our camp.   We voted them
unanimous thanks; and, placing a guard, tumbled
down into our nest.   But slumber came not,
though the morning did, when we expeditiously
roasted and ate our turkey, and sailed away
down stream, giving vengeful shots at the roaring
monsters, who were now heard in all directions
bellowing like bulls.   This noise they frequently
make in the early morning of a warm day; and,
to those first experiencing it, it strikes alarm : but
this soon passes away.

We steam on, and past what were, a hundred
years ago, plantations of indigo and cane, owned
and cultivated by the Spaniards; past sites of

cities and villages of the aboriginal races, where now the tall pine and bright-leaved magnolia and sweet-bay flourish. Then past the pleasant shores and the extending savannas, half submerged, and dotted all over with sleeping islands; then by the quaint capes and headlands, all glorified by the rising, cloudless sun, our thoughts alternately enrapt by the gushing songs of the morning birds, and the remembrance of the tide of human life that swept the bosom of this river centuries ago, where now none but a few curious and idle visitors sail pensively over its waters.

CHAPTER XI.

HAVING returned to Picolata from the charming
scenes and wild adventures of the upper and the
incomparable river, we now, with a dozen of fellow-
travelers, turn our faces eastward toward the
ocean and to St. Augustine. There are two ways
of reaching this, the oldest city in America, and
the Mecca of all Florida tourists. One is by the
steamer " Henry Burden," from Jacksonville,
down the river and round the sea-coast, — a run
of perhaps eight hours ; sometimes a smooth pas-

150

sage, and again a somewhat rough one. To those who do not dislike a sea-voyage, this is a very pleasant trip. The price of passage is four dollars each way. Only weekly trips are made, leaving Jacksonville in the latter part of the week and returning the following day.

The other route is to take a river-boat, — and they are to be had nearly every day, or every other day, at Jacksonville, — and proceed up the river to Picolata, then across the country eighteen miles by stage-line. The fare this way is two dollars on the steamer, and three dollars from Picolata by stage. The fare by the outside line is reasonable, since the steamer runs to accommodate that particular travel; while by river-boats it ought not to exceed one-half their present rates, as these boats do not run on the river solely for this travel. Then, again, the stage-fare is fully one-third in excess of paying-rates.*

---

* Since writing the above, we are informed that the St. John's Railroad is being built, and will run between St. Augustine and Tocoi, on the river, forty miles above Jacksonville, in time to accommodate the travel of the coming season. Of couse, this will then be the route for travelers proceeding to St. Augustine, *via* Jacksonville, thence by steamer to Tocoi, thence by rail across.

Picolata, as we have remarked in a preceding chapter, is simply a settlement of one family — nothing more. If the reader were to take passage on the Pacific Railway, and get off at the last new made and named station anywhere along its entire length, he could not well be left more in the woods than travelers are left at this *town* of Picolata. But what of its history? Next perhaps to St. Augustine itself, it was among the earliest founded of all the Spanish settlements. Two hundred and fifty years ago, there were more white people stationed and trading here than we find to-day. It grew to be in a hundred years, under the Spanish rule, a town of more than a hundred dwellings and shops. It was the main artery of supply for the up-country, and chief channel for the return of all articles of export, including furs, indigo, sugar, and fruit. Crossing over to St. Augustine, these were sent in trading-vessels home to Spain. The order of Franciscans erected here an imposing church edifice; and within its walls assembled the devotees of Spain and the Pope, together with the proselyted savage, still plumed and decked in his wild costume. But, of all this ancient civilization, scarce a vestige remains: by savage wars and the desolation and destruction of towns and missions and plantations, together with the changes in government, the transformation has been made complete; and now the original wilderness everywhere covers the State, and as nearly primeval as in the time of Adam.

Opposite Picolata, on the western shore, is a mark of these olden times that has however withstood the leveling influence of the past. It is a great earthwork fort, with its sharp angles smoothed and round, yet still evidencing its original strength and power. Its walls must have been twenty feet in heighth, and were constructed entirely of the surrounding earth. It is now overgrown with grand old oaks, deep rooted in its walls, and the wild birds and beasts have made here their home.

Those who have been beguiled into reading the small volume on St. Augustine, said to have been written by Mrs. Yelverton, will be doomed to a bitter disappointment in the plain, and somewhat unromantic and uninteresting stage-ride, which she, in her volume, has made so marvelously enchanting. She exhausts the vocabulary of flowers and sentiment quite, in her florid delinea-

tion of what the country is, or was. We found
this ride of eighteen miles one of the most disa-
greeable and miserable we ever made, and
through a tract of country entirely devoid of
unusual attractions.

After a protracted delay, the stage and horses
were ready; and such Rozinantes as they were,
too, — just fit to be knocked in the head by a
windmill, and so starved-looking as to suggest
the thought that they were the very horses which
are said, after considerable experimenting, to have
been successfully taught to live without eating.
The drivers seemed a cross between the lazaroni
of Italy and the village loafer. Our party was
too large to be all carried in a single coach, and
an old rheumatic wagon, without seats or covering,
was tied up, and braced with sticks, for the
journey. The trunks served for seats; and, thus
equipped, we set out upon our winding way. It
was five o'clock in the afternoon when we started ;
and at half-past ten that night, we struck the
ferry, just outside the old walls of St. Augustine.
The reader can imagine the speed we came the
eighteen miles. The prospect bade fair at one
time for us to camp among the pines, along the

roadside. Our horses, " Duroc " and " Fire-fly," were no " Dexters." They had, as we have observed, learned to live without eating; but, hauling a wagon eighteen miles through the sand, they had not calculated upon, in connection with their abstemious habits. Their wind gave out; and, having no special knowledge of time or tune, they got off irregularly, first " Duroc," and then " Fire-fly," shot a neck ahead. The lumbering coach behind us fared no better. Their four horses were reduced to three, one having been turned out to *grass*, we were about to say, but that would be an imposition on the dumb beast, and we have the fear of Bergh before our eyes, — it was *out to sand*, Spanish bayonets and pine-trees. For our part, we should have preferred the sand, of the three, as that did appear white and almost palatable, even to ourselves; for we had tasted nothing since noon, and it was fast approaching midnight, to say nothing of the appetizing effects of our eighteen-mile ride in a lumber-wagon with trunks for seats ! In our younger days we have read of the lonely traveler on the desert, who afar off descried a faint light that cheered him on, and brought him safely to

shelter and to rest. We think all had a lively
sense of appreciation of this traveler's gratitude,
when we hailed the light of the ferryman's fire
burning on the roadside in front of his cabin.
Even the poor jaded horses took courage, and
absolutely dashed forward at an unaccountable
rate. Across the St. Sebastian River, on the
rope-ferry, and a rapid drive through the dark
and narrow streets, and we were ushered into the
Florida House, tired, and hungry, and battered.
A plain, comfortable, clean room would have
been a luxury: it was what we had hoped for, a
place to lay our bones for a refreshing sleep; but
this was not on the programme. We were told
the house was filled, and the town crowded, but
we could have a room. Dispatching a hurried
meal, we were shown through the house and
across the yard, up a rickety pair of stairs, to a
filthy room over the dirty kitchen, the former
"negro quarters," and with a stinking black lamp,
that had been innocent of soap and water since it
fell into its proprietor's possession. Left here in
this indecent room, with a flavor of onions and
stews from below, and a sickening odor of bugs
above, our first impressions of St. Augustine were

likely to have a rather biased and unpleasant coloring. Determined not to be considered a *growler*, and feeling too stupid to become properly enraged, we concluded to stay in that room, *and we did.*

A comfortable breakfast improved our condition; and, not wishing to cultivate the further acquaintance of the lazy and incompetent young man who was the only landlord we saw, we left for private lodgings.

We have in the historical review of Florida, in the earlier chapters, given a comparatively full and accurate account of the settlement and history of this ancient town. To these chapters the reader is referred for historical details. More than half a century prior to the landing of the Plymouth colony, this was a town of considerable reputation, and long before 1620 was a fortified stronghold, challenging the stoutest navies on the globe.

The appearance of St. Augustine to the visitor is as quaint and peculiar as its history is bloody and remarkable. Nothing like it is seen in this country; and, having been built by a people so entirely different in manners and customs from our own, it has been surrounded with an interest not shared by any other city in the land.

For a sketch of this old city, we prefer to give
that written by Rev. H. Clay Trumbull, in " The
Springfield Republican," though we can hardly
share all his sentiments regarding this antiquated
and dilapidated town.

" Its principal building material is a unique
conglomerate of fine shells and sand, known as
coquina rock, found in large quantities on Anastasia Island, at the entrance of the harbor, and
which is easily cut in blocks to be laid in courses,
and perhaps covered over with stucco. The
streets are quite narrow ; one, which is nearly a
mile long, being but fifteen feet wide, and that on
which a principal hotel stands being but twelve
feet, while the widest of all is but twenty-five
feet. An advantage of these narrow streets in
this warm climate is, that they give shade, and
increase the draft of air through them as through
a flue. Indeed, some of the streets seem almost
like a flue, rather than an open way ; for many
of the houses, with high roof and dormer win-
dows, have hanging balconies along their second
story, which seem almost to touch each other
over the narrow street ; and the families sitting
in these of a warm evening can chat confidential-

STREET VIEW IN ST. AUGUSTINE.

ly, or even shake hands with their over-the-way neighbors.

"The street walls of the houses are frequently extended in front of the side garden, — the house roof, and perhaps a side balcony, covering this extension; or the houses are built around uncovered courts, so that, passing through the main door of a building, you find yourself still in the open air, instead of within the dwelling. These high and solid garden walls are quite common along the principal streets; and an occasional latticed door gives you a peep into the attractive area beyond the massive structure, with perhaps a show of huge stone arches, or of a winding staircase between heavy stone columns, or of a profusion of tropical vegetation in the winter garden, bringing to mind the stories in poem and romance of the loves of Spanish damsels, and of stolen interviews at the garden gate, or elopements by means of the false key or the bribed porter. The principal streets were formerly well paved or floored with shell concrete, portions of which are still to be seen above the shifting sand; and this flooring was so carefully swept, that the dark-eyed maidens of old Castile, who then led in society

here, could pass and repass without soiling their
satin slippers. No rumbling wheels were permitted to crush the firm road-bed, or to whirl the
dust into the airy verandas, where, in undis-
turbed repose, sat the indolent Spanish dons and
dames.

"Built as a military town, the city was formerly
walled across its northern end; which sufficiently
protected it, as it stands on a peninsula nearly
surrounded by the St. Sebastian River and St.
Augustine Bay. The gateway of the old wall
still stands, and is quite an imposing ruin, with
ornamented lofty towers and loopholed sentry-
boxes. The ditch before the old wall (or possibly it was a stockade, except at the gateways) is
clearly marked, and even yet partially filled at
high tides. It runs from shore to shore, and was
evidently broad and deep. The old fort, once
called San Juan, then St. Marco, but now known as
Fort Marion, is a curiosity. It stands on the sea-
front, at the upper end of the town, the wall or
stockade formerly running from it to the gateway,
and west to the river. Its material is the inevitable coquina rock. It was a hundred years in
building. While owned by the British, it was said

to be the " prettiest fort in the king's dominions."
Its castellated battlements; its formidable bastions,
with their frowning guns; its lofty and imposing
sally-port, surrounded by the royal Spanish arms;
its portcullis, moat, drawbridge; its circular and
ornate sentry-boxes at each principal parapet-angle,
its commanding look-out tower; and its stained and
moss-grown massive walls, — impress the external
observer as a relic of the distant past; while a
ramble through its heavy casemates; its crumbling
Romish chapel, with elaborate portico and inner
altar and holy-water niches; its dark passages,
gloomy vaults, and more recently discovered
dungeons, — brings you to ready credence of its
many traditions of inquisitorial tortures, of decaying skeletons found in the latest-opened chambers
chained to the rusty ringbolts, and of alleged
subterranean passages to the neighboring convent.

"These stories lose none of their force by being
recited in the fitful light of the dim lamp of your
military guide, as you follow him into the damp
and noisome recesses to the echo of your own foot-
fall or the grating lock and creaking hinge of the
slow-swinging ancient doors. Many a dark tally-
list on the moldering walls, or a rudely-executed

11

sketch, shows how the dragging days were noted or employed by weary prisoners of long ago ; and the narrow loopholes are shown through which the two Seminole chiefs attempted their escape, one making it good, and the other sticking fast in the crevice until he was rescued with barely his life remaining. At the time of Gen. Oglethorpe's attack on St. Augustine, the old fort, or castle as it was then called, stood a bombardment of thirty-eight days from batteries erected on Anastasia Island. But the injury to the fort was only slight ; for the spongy walls of coquina received and imbedded the heavy shot, as would the embankment of a modern earthwork. The marks left by the shot are plainly seen to-day. But time is at length doing its work with the old fort. Its walls are showing huge fissures, and on recent inspection it was declared unfit for further defensive service.

" In the buildings of the town are some remains of elegance, as well as much of antiquity. The cathedral is unique, with its belfry in the form of a section of a bell-shaped pyramid, its chime of four bells in separate niches, and its clock, together forming a cross. The oldest of these bells is marked 1682. The old convent of St. Mary's is a suggestive relic of the days of papal rule. The new convent is a tasteful building of the ancient coquina. The United-States barracks, recently remodeled and improved, are said to have been built as a convent or monastery. The old government house, or palace, is now in use as the post-office and United-States court-rooms. At its rear is a well-preserved relic of what seems to have been a fortification to protect the town from an over-the-river or inland attack. An older house than this, formerly occupied by the attorney-general, was pulled down a few years ago. Its ruins are still a curiosity, and are called (though incorrectly) the governor's house.

"The 'Plaza de la Constitution' is a fine public square in the center of the town, on which stand the ancient markets, and which is faced by the cathedral, the old palace, the convent, a modern Episcopal church, and other fine structures. In the center of the plaza stands a monument erected in honor of the Spanish Liberal Constitution. When the Constitution was abolished, these monuments in all dominions of the crown were to be destroyed; but a compromise was effected on this by the removal of the inscribed tablets. On the cession of Florida to the United States, the long-concealed tablets were brought from their hiding-places, and re-inserted in the monument. On this plaza were burned effigies of John Hancock and Samuel Adams, early in our Revolution, while the British held Florida.

" The old Huguenot burying-ground is a spot of much interest; so is the military burying-ground where rest the remains of those who fell near here during the prolonged Seminole war. Under three pyramids of coquina, stuccoed and whitened, are the ashes of Major Dade and one hundred and seven men of his command, who were massacred by Osceola and his band. A fine sea-wall of nearly a mile in length, built of coquina with a coping of granite, protects the entire ocean front of the city, and furnishes a delightful promenade of a moonlight evening. In full view of this is the old light-house on Anastasia Island, built more than a century ago, and now surmounted with a fine revolving lantern.

" The street names, Cuna, St. Hypolita, Tolomato, St. George's and the like, have an ancient and a foreign smack about them; while the family names, such as Dumas, Fatio, Hernandez, Oliverez, Alveres, Monardi, Segui, Andrea, Sanchez, Medices, and Bravo, mark it as any thing but American in its origin. Some of the Roman-Catholic customs of carnival and evening serenades before Easter are still kept up by the Minorcan population."

A word as to these people, who constitute no inconsiderable portion of the present population of St. Augustine. While Florida was in possession of the English, a Dr. Turnbull went to Greece, and received permission to transport such families as chose to go to Florida. Obtaining a small number, not enough for his proposed colony, he halted at the islands of Corsica and Minorca in the Mediterranean, where over a thousand joined his company. They landed just inside of Musquito Inlet, at New Smyrna, some seventy-five miles south of St. Augustine. Turnbull soon became imperious, and by the aid of a few immediate friends reduced these patient, hard-working people to a state of slavery, assigning them tasks under overseers, and treating them in the most shameful manner. His promises of lands and creature comforts, made at the time of their joining his expedition, were disregarded, and with acquired wealth came added austerity and hardships for these now dependent people. Thus for nine years they were in bondage, when, stung to resistance, they assembled clandestinely, and marched in a body to St. Augustine, where they were kindly received, and allowed to remain. They form a very quiet class, attentive to their own affairs, and never meddling with their neighbors. They are intelligent and industrious, and some have acquired considerable property.

There are a few fine residences in St. Augustine; and these, with their ample surroundings and beautiful gardens, give a heightened interest to the place. Senator Gilbert has a summer residence here, the first as you enter the town, by the bridge, on the right; then Buckingham Smith's, nearly opposite, and Dr. Bronson's on the plaza, with others, are beautiful homes. A profusion of tropical plants and shrubs and trees ornament their grounds. Here the orange flourishes, and is abundant and delicious: several fine groves invite the visitor's inspection. The fig and date and palm and banana are all seen here, as also the lime and lemon, which grow to a great size, and the sweet and the wild olive; the citron, the guava (from which a delicious jelly is made), and the pomegranate, are all indigenous. This is the home of the grape, and peaches luxuriate in this climate, as likewise the Japan plum.

Besides the gardens spoken of, we see few flowers; and this is what quite astonishes us in this "land of flowers," where they grow so easily, and with so little care that there seems no excuse why *all* the gardens should not have these simple yet beautiful adornings.

For many years the town has been at a stand still, and property at a low figure. Good titles can with difficulty be obtained; and this is now the great drawback to the improvement of the place, though within a few years Northern people have been coming in and taking such titles as were offered. One gentleman, Mr. Howard, from New York, has within a year past invested near fifty thousand dollars in real estate in the city, which is beginning to feel the effects of this healthful influx, property having already risen to fourfold its value five years ago, and still not high. The residence of Senator Gilbert, before alluded to, was bought by him at the close of the war, as we are informed, for about eight thousand dollars, and we judge worth forty now. This place has several acres of ground in it.

There is one thing the town lacks, and that is, a comfortable hotel. This it has not at present, though a more splendid opportunity never seemed to present itself. Hundreds visit this city during the season, and other hundreds are kept from coming in consequence of this lack. Our own experience attests the great want. There are three hotels here, so called; but a respectable mechanic's and clerk's boarding-house at the North is, in most respects, their superior. The Florida House — well, it don't deserve either patronage or mention, except that it is now for sale; the owners, not unlikely, having become themselves disgusted with it. The place for strangers is the private boarding-house, of which there are, luckily, quite a number; and they seem so well patronized, that, to secure accommodations for any thing of a party, it is needful to make application a week before your arrival. Mrs. Abbott's, Mrs. Gardner's, Madam Fatio's, and Mrs. Dummet's are among the best known, and are all pleasant homes, and furnish excellent accommodations, at about half the price of the hotels. There are a half-dozen other private families, whose names do not so readily occur to us, but who are well known to the stage-lines, where visitors are quietly provided for and made comfortable.

But, as we said, the great need is a hotel, not elegant nor expensive, but convenient in location and arrangement, where the invalid and tourist may find a home. The season commences early in November, and ends in May, — full seven months in which to reap a harvest ; and some remain the year round, finding the climate here less trying in summer than at the North. Every third man you meet from the North is arranging to build a hotel ; but since they do not grow of a night, like Jonah's gourd, why, St. Augustine will have to wait till one is built in the usual way. The last party who had shouldered the undertaking was a Baltimorian, with a very Yankee countenance ; but what will come of it time will tell. Burns has written something in relation to " the best-laid plans of mice and men," &c. ; and, as we reflect, we have some doubts of the result of even this last project. If, however, St. Augustine wishes to become prosperous, and her citizens to accumulate of this world's stores and to secure the annual

influx of visitors, a new hotel is the first stepping-
stone, and can not be lost sight of. The people of
the place could afford to levy a tax for its erection,
rather than it should go undone. This being accomplished, with a railway from the city to Picolata, which was once graded, the tide of travel will
be greatly increased.*

The longer one remains in this antique town,
the more he is attached to it: at least, this was
our experience. It improves on acquaintance.
The plaza, or public square, affords a pleasant
retreat from the sand, which everywhere else
covers the place. Here are shade-trees, and the
firm green turf and benches, whereon the visitor
may lounge, and idle away the hours. At the foot
of the square, which fronts on the bay, is the
market-house, so entirely different from those elsewhere seen; being here neat, airy, and attractive.
It consists of a roof supported by brick pillars, a
half-dozen on either side, with a floor of the same
material, and is altogether unique in appearance.

The military band, on pleasant evenings, has

* Since penning the above, we have been informed that the work of
grading and track-laying on this proposed railroad has commenced with
a prospect of its being finished this year. Its terminus, on the St. John's,
is Tocoi.

been accustomed to assemble on this square, and discourse the national airs and other melodies, to the great delight of the people. This source of pleasure and amusement has ended, we regret to say, in consequence of a recent act of Congress, consolidating regiments, and reducing the number of military bands in the service.

The number of strangers here greatly exceeded our expectations, and thronged in every street and public place. The fashionable belle of Newport and Saratoga, and the pale, thoughtful, and furloughed clergyman of New England, were at all points encountered. The meeting of friends whom we had not seen for years, and others whom we had never met, but yet could call our name, seemed strange and quite a dream.

Our visit being ended, and a few photographs secured of points of interest as mementoes, together with a cane of the pomegranate, presented by a friend, which he had cut from his garden, we commenced our journey back over the road we came, to Picolata. It was a bright April morning; and as we sped out of the city and over the St. Sebastian River, through the thickets, and into the piny woods, a pang was felt at parting from the old city and our new-made friends. We had now the day before us, with new horses, a lighter carriage, and a gayer company ; and what before was a tedious, wearing ride, was now made a pleasing journey of three hours, though among the inevitable pines, only here and there broken with clumps of clustering vines and overarching branches, whose friendly presence was owing to swampy places by which the roadway led.

At Picolata we again hailed the river, where a steamer was in waiting to bear us on our journey.

CHAPTER XII.

THE CLIMATE.

THE chief interest, perhaps, that the State of Florida possesses for the people of the North, is its delightful climate, and the reputed beneficial effects thereof on the health of certain classes of invalids.

Its location is the most southern of any portion of the United States, and is the most tropical in character. Reaching almost to the tropical zone, and extending up to the thirty-first degree of latitude, its entire coast-lines are bathed by the warm waters of the surrounding seas; while the gentle trade-winds cool and purify its atmosphere, making the peninsula, as a place of residence, both healthful and delightful.

Invalids have long sought this portion of the Union; and its general reputation has steadily increased, till now scores and hundreds annually migrate to some point in the State, as their predilections seem to favor : but as a rule a majority of them remain upon the St. John's River and its tributaries, or else upon the Atlantic coast at St. Augustine and the Indian-river country, which is an extensive inlet running very near and quite parallel to the seacoast at the central portion. The story was everywhere current along the river, that full fifty thousand people had, this last season, visited Florida. This, of course, included all classes : but we can scarce credit so large a statement; and if we cut it down one-half, then the statement may be taken with some allowance for interested motives. There is no denying, however, that great numbers have visited the State within the winter 1868 and '69. The chief hotel in Charleston, S. C., the Charleston House, was kept crowded to its utmost capacity during the winter by this Florida travel. It may not be too much to say, that nine-tenths of all the arrivals at that house were on their way to or from that State. We mention this as indicative of the growing and already great importance of the question as to the effects of the climate upon invalids and those who seek rest and recuperation from the steady and exacting demands of business. There is needed among those who fill the various professions more of rest and play than they get.

While we write, an eminent gentleman connected with one of the chief journals of the metropolis has been suddenly snatched away, in the full meridian of life, from over brain-work. To say that his was an impaired constitution would not be warranted, since he had a perfect physical organization, and, mentally, was as well balanced as any man we ever knew. Mr. Raymond possessed an intellect of the very highest order, and practical talents exceeded by but few in the history of our leading men, together with a capacity for steady mental exertion beyond all others; and yet this noble man fell a victim to over-work. His health was indeed so perfect as not to break, as in most cases would be the result of such intense application, and thus sound a warning. It is this over-worked class, as well as the invalid, who need to go to Florida. For the former, it is just the place in which to spend a winter : there is no doubt in our minds of the benefit they would derive by a few month's residence in that climate. Then there are the weak and nervous, — those whom care and anxiety have broken down, and who now need, above all things, a change of scene, and a quiet life away from their former surroundings. With a new diet and a *dozing* climate, they will rapidly recover : these, therefore, need not hesitate to pack their trunks and start for Florida. We know of no place equal to it for persons thus afflicted : all improve under the influence of this warm and genial climate, where a comparatively even temperature is maintained, and where the rule is cool nights, in which sleep, the sweet restorer, comes with so many blessings to the fevered and fretful invalid, and the over-worked. No physician is so skillful, or remedy so marvelous in restorative power, as sleep. This the resident of Florida may more easily obtain than in any other climate of which we have any knowledge. The winters are not so cold as to freeze during the night, or to necessitate artificial, over-heated air in the dwellings, rendered often quite impure by this furnace-system so general in the North during the cold season; nor are the summers so sultry and heated as to deprive you of rest, as is the case in the severest hot weather in nearly every other portion of the country. The thermometer never settles as low, or rises as high, as at any point between this State and Canada. The lowest point reached in winter is seldom below thirty degrees, while in midsummer it rarely exceeds ninety-five degrees; the average being, for the three summer months, about eighty degrees. In New York, Boston, or Montreal, every summer carries the thermometer to a greater height. The earliest frost recorded occurred on the 27th October, in 1857; and the latest frost was in February (the 14th), 1859. Severe frosts usually occur in January, when ice is formed in pools of water or buckets, if left exposed. Since 1835, no very destructive periods of cold weather have been experienced; then it was *cold*. People who were living in the State at that time speak of it as a severe cold snap, reminding them of Northern latitudes. The vines and shrubs and orange-trees, with many other kinds of trees, were quite annihilated; and what are now seen have been either grown from

12

the seed, or are sprouts from the old stumps of the frost-killed trees.

It is the severe and sudden changes in temperature at the North that do the injury to enfeebled constitutions. One day in winter it is quite mild and pleasant; while the next morning it is so dreadfully cold, that the going out of doors is a trial to the able-bodied, and a severe shock to those lacking in vitality. The spring is even worse than the winter ; for while the latter, though cold, has a dry atmosphere for the most part, the former is piercing, cold, and wet, and miserably coquetting, with all degrees of temperature in a single day. Spring has in fact got to be, if, indeed, it has not always been considered in most of the Northern States, worse for all kinds of invalids than any other season of the year; and where the east winds prevail at this season, the mortality list exceeds, for March, April, and May, that of all the balance of the year. An escape from these months to a more equable climate is, to the invalid afflicted with pulmonary difficulties, a vital one.

A continuous, steady cold, dry climate, or an even warm one, is the most to be desired for a majority of the suffering and afflicted of our race. Florida and Minnesota are the two points which most nearly represent these conditions. They have been frequently contrasted, and, like every other subject, each has its special advocates. We shall not even pretend to decide between them, for undoubtedly both are beneficial as a resort; but to determine which of the two is the better adapted to benefit certain cases is beyond possibility, since they differ in many particulars: that is for the family physician to decide, who should know the constitution and habits of his patient, and whose counsel should always be weighed in all matters of this kind. Only general hints can be given, and each is to determine for himself between the one and the other, or whether remaining quietly at home may not be best. When patients are so debilitated as to make traveling a trial and a burthen, they should remain at home, where their nearest friends may watch and tend them; but in the incipient stages of tubercular formation, with a judicious change of residence and a nutritious diet, coupled with great care, the disease may be arrested, especially if resort is had to gentle exer-

cise in the open air. Northern latitudes admit
of exposure to the weather only during the summer season, and herein lies the great advantage
of a residence in Florida. We met at St. Augustine a lady from Syracuse, afflicted with pulmonary disease : we should judge it was constitu-
tional in her case. She stated she had been
unable to go out of doors during the cold and wet
weather of the preceding winter, at her home,
but had not failed to walk out daily (except it
was raining) during the whole winter in Florida.
She had at first, in coming into the State, spent
several weeks on the St. John's River, and then
took up her residence at St. Augustine. This
was undoubtedly a very judicious plan ; for the
climate of the river differs materially from that of
St. Augustine, on the seacoast. The former is
milder and more gentle ; and the patient suffering
with disease of the lungs would do well to remain
on the river for a while, and then the climate of
St. Augustine, with its sea-breeze acting as a mild
tonic, braces up the system.

The whole peninsula is in the range of the
trade-winds, and is swept by them daily, ren-
dering it as cool and pleasant as one could ask.

Sufferers from nervous prostration and general
debility need not delay their visit to St. Augustine for any reason above given, since they would
probably experience as great benefit at the outset
as at any subsequent period of their sojourn.

People usually do not go to Florida before early
in November, though they might leave home at an
earlier date, making tarries on their way; but, if
they reach the State by that time, they will not
have gone wrong. They can remain until the
first of May, when it is safe to return to the
Northern States.

We speak in reference to invalids, of course.
Pleasure-seekers, or those in health, may visit
Florida, and the next week take a sleigh-ride
among the hills of New England with impunity,
perhaps. There are those from the North, both
invalids and others, who make Florida their home
the year round; and they speak in the very highest
praise of the climate during the summer, declaring
they do not suffer with the heat as much as they
formerly did at home, and that the benefit derived
from a residence is increased by remaining. This
may be true in some cases, whereas it might be too
debilitating in others; not from the great rise in

the thermometer, but from the long continuation of
hot weather. It commences to be warm in April.
We have seen the thermometer in the early portion of that month as high as eighty degrees in the
shade at two, P.M. True, that was exceptionable ;
but it was warm weather at mid-day through the
half of that month ; and to continue this temperature on to October makes a long season of summer weather, which might be objectionable in
some instances.

There are those who assert the climate of
Florida to be proved, from statistics, to stand at the
lowest rates of mortality. From the census report
of 1860, we find the average number of deaths
from consumption, in various States, to be —

> One in   254   in Massachusetts.
> One in   473   in New York.
> One in   757   in Virginia.
> One in 1139   in Minnesota.
> One in 1447   in Florida.

This table certainly speaks very highly in
favor of the climate for this class of diseases.
It is not impossible, however, that the returns
may have been less perfect for Florida than those
of other States, owing to the sparseness of popula-

tion and the inferior facilities for obtaining exact data.

The following table of observations made at Jacksonville shows the highest and lowest range of the thermometer each month for five years from 1857 to 1861, both inclusive : * —

MONTHS.	1857		1858		1859		1860		1861		REMARKS.
	H.	L.	H.	L.	H.	L.	H.	L.	H.	L.	
January...	72	16	76	38	76	30	76	40	..	..	Ice one to two inches
February..	81	44	77	39	79	39	79	44	75	42	thick, Jan. 19, 20, 1857.
March	85	41	83	34	84	45	83	40	83	43	
April	81	47	86	49	89	53	92	58	85	54	
May......	91	61	91	66	92	64	92	58	94	64	
June......	91	73	92	73	94	70	97	69	98	73	
July	89	68	96	74	95	70	98	74	92	70	
August ...	95	75	94	75	91	75	93	73	91	73	
September.	92	64	86	64	92	70	89	65	92	58	
October ...	81	42	85	62	84	50	87	53	86	57	At 7, A.M., Nov. 25,
November .	82	27	79	39	79	35	80	25	79	45	1860, the thermometer
December .	80	39	78	40	79	36	72	32	74	38	stood at 25°.

There is a marked difference in the thermometric range at Enterprise, two hundred miles south of Jacksonville; not that it so much exceeds, if any, the above table, but that it does not reach as low a point, making the climate more even in temperature, and consequently more desirable. This,

* This table is from the report by the Hon. J. S. Adams, the Commissioner of Immigration, published in 1869.

even, is observed at Palatka, which is also a more favorable place for these reasons.

A comparative table, from the same source, showing the monthly and yearly mean for twenty years at St. Augustine, thirty-one years at West Point, and thirty-five years at Fort Snelling, is given, showing the equability of climate of the several localities : —

	Jan.	Feb.	Mar.	April.	May.	June.	
St. Augustine, Fla..	57.03	59.94	63.34	68.78	73.50	79.36	.....
West Point, N.Y. ..	28.28	28.80	37.63	48.70	59.82	69.41	.....
Fort Snelling, Minn.	13.76	17.57	31.41	56.34	58.97	68.46	.....

	July.	Aug.	Sept.	Oct.	Nov.	Dec.	Year.
St. Augustine, Fla..	80.90	80.56	78.60	71.88	64.12	57.26	69.61
West Point, N.Y. ..	73.75	71.83	64.31	53.04	42.23	31.98	50.73
Fort Snelling, Minn.	73.40	70.05	58.86	47.15	31.67	16.89	46.54

None need expect that every winter day in Florida will be like a selected day in May or September at the North. There will be cool and cloudy days; there will be occasional rainy days, though the winter months are usually very free from rain. The rainy season is in the summer; and of these months August is usually prolific in heavy falls of water. In most countries where they have what is denominated the wet season, this occurs in the winter or spring months, leaving the summers, when vegetation needs rain most, very dry, and trying to the crops; whereas in Florida this is reversed, and in the hot weather the heaviest rains fall.

The mild atmosphere of winter, which permits so much life in the open air; the sea-breezes from the ocean on the one side and the gulf on the other; the mode of living without air-tight stoves and hot-air furnaces, but with ample ventilation in, around, and under their dwellings, which have no cellars, and usually stand on posts a couple of feet from the surface, — to these things may be ascribed the freedom from lung complaints, though the character of the soil itself greatly contributes to the absence of this disease, by its loose, sandy nature quickly absorbing moisture, and yet being most of the time warm and comparatively dry.

Visitors may expect to find the evenings and nights often damp and chilly in winter, and occasionally frosty. Our May and September have such nights; and yet these months are thought to possess the most agreeable temperature in the North. Snow is not seen, nor ice, except as rarities. It is no unusual thing to sit without a fire and with open doors in the winter; and it is never absolutely uncomfortable to be out of doors, except in a storm or at night. Fires are, however, a great comfort most of the winter evenings and mornings, chiefly in removing the dampness, and taking off the chill; but an open door or window at the same time is not a discomfort. A good idea of a Florida winter may be best had, perhaps, by letting our months of September, November, and May, stand for their December, January, and February.

There is no doubt, notwithstanding the fine climate, that people do die in Florida! They have hitherto, and will continue to drop away there as elsewhere from some one of the many ills that all flesh is heir to. Fevers, malarious and bilious, and fevers with chills, are not infrequent, as they will always be in a country that is new, and heavily timbered, and filled with rivers and numerous lakes. Still, we believe that the ague prevails more in the Wabash Valley, and that fevers were as common in many other States, where now a good degree of health is maintained, in their early settlement, as at present in Florida.

Of course, the climate is not a panacea for all ills. All pain will not vanish under its influence, nor will it act as a certain balm to those whose disease is deeply seated and far advanced. Alluring hope should not tempt such from the comforts of home, even for the placid skies of Italy. There is no question that many who are constitutionally predisposed to, and those who are in the first stages of, pulmonic troubles and kindred complaints, also those wanting in vitality, and the over-worked, and those suffering from nervous prostration, may reasonably hope for the most beneficial results; but all should exercise the same care here as at home. As a rule, they should not expose themselves to the night air, nor be tempted on warm, bright days to lay aside thick shoes and comfortable clothing. The invalid should always be clad in woolen clothing; and the robust do not require a linen suit except in the summer months. Better suffer with the heat at mid-day than with a feeling of chilliness at sunset.

CHAPTER XIII.

THE SOIL.

First Impressions. — Varieties of Soil. — Cane, Cotton, and Corn. — Adaptability of the Soil for Early Vegetables. — The Culture of Rice, Coffee, and Tea, considered. — The Chinese. — A Prediction. — Good News to all Housekeepers. — A California Senator. — His Triangular Position.

THE soil of Florida differs very greatly from that of every other State with which we are familiar. Its first appearance is that of a worthless character, unproductive and quite valueless; but yet you are everywhere confronted with such stately, vigorous trees, and rank growths, that it seems quite paradoxical that out of the sand should come so giant a Flora. On closer inspection, we find the subsoil is often a mold, and, again, filled with broken marine shells and fragments of lime-rock ; and this has worked itself in with the fine sand, giving it its power to sustain such forests as abound everywhere along the river-courses. Over the whole State there are, of course, great varieties of soil. In the rich hummock and bottom lands, a black mold, with a mixture of the sand, is seen; the dark, vegetable compound predominating: and you have this soil shaded off till you come to the almost pure silica, or dry sand, where little else but the scrub-oak and black-jack flourish. The pine-forests, as we have before remarked, cover a majority of the surface in the northern and eastern half of the peninsula. These pine-lands are not so desirable for crops as the river-lands; yet they, even, are valuable. The first-rate pine-land possesses considerable vegetable matter, and has a kind of marl or limestone substratum, that gives it an enduring fertility. This sort of land is, perhaps, to the early settler, the very best, since he may always find within the pine-woods a suitable and healthy location for a dwelling. This is not always true of the river-banks, where the hummock-lands are. Time, and the opening-up of the country, will render them quite habitable; but, at present, chills and fevers have to be encountered unless the site is very high and airy. The settler may, however, by judicious purchase, procure on almost any portion of the St. John's River a fifty or a hundred acre tract,

which shall not only front on the river, embracing
considerable of the rich river-land, but extend
back into the pine-woods, where he may locate his
dwelling, and be tolerably free from the diseases
which come from a home near the lowland and
standing waters of the bays that indent the shores.

Much of the second-class pine-land will grow
nearly all kinds of vegetables; and even cotton
and the cane will flourish, this having a clay and
marl foundation. Three hundred pounds of *Sea-
Island* cotton have been taken from an acre; and
Cuba tobacco and all kinds of fruit may be cultivated with success upon it.

It is the bottom and river lands, however, that
seem the most highly prized; and this is not
strange when their productive power is so great.
The sugar-cane *matures* in this country only in
Florida. True, it is grown in Louisiana and Texas
extensively; but it no where tassels out like our
Northern corn, except in this peninsula; and this
is the sure sign of its maturity. The term
" hummock " land — the Indian name for elevated
tracts lying above the low or wet lands — is now
applied to nearly all the hardwood lands in the
State. It may be cultivated, year after year,

VIEW ON LAKE HARNEY.

with the most exhausting crops of tobacco or
cane, without apparent diminution of its power.
For gardening purposes it is unsurpassed; and is,
indeed, what the greenhouse man sifts and mixes
in fine compost, for his delicate plants and shrubs.
The time is coming, when, with a semi-weekly
line of steamers running direct from Jacksonville
to New York, the markets of that city will be
supplied with early vegetables grown on the St.
John's River. A few industrious Yankees have
already initiated the enterprise of early gardening,
and there is not a shadow of doubt as to the
result. Green pease may be had in abundance by
the 1st of April; and farther up the river, new
potatoes can be grown for market by that time.
We know this to be possible; for they were upon
our table as early as that, as were also straw-
berries and blackberries. The ordinary vegetables can be raised to great profit for the Northern
markets: pease, potatoes, melons of all kinds,
string-beans, &c., mature rapidly and early,
and with very little care beyond planting and a
single hoeing. Of all crops, the melon grows the
most spontaneous, and may be raised on the very
poorest soil, as light and air contribute chiefly to

their growth. All these articles can be safely shipped, even with the present means of conveyance to New York. Sweet potatoes are, of course, grown here, and they sell at an extremely low rate. In this article alone a handsome profit might be realized. By and by, when quick water communication is had, tomatoes, cucumbers, squashes, and other articles, may be sent to distant markets : at present these products, embracing all we have named, except, perhaps, sweet potatoes, find a ready sale at Jacksonville, as the demand exceeds the supply.

Under British occupation, there were extensive indigo plantations, this crop forming the chief article of export ; and even now it may be seen growing in its wild state, having become naturalized and indigenous. Coffee and rice, especially the latter, could easily be made profitable ; and their culture will some day employ, doubtless, many hundred pairs of hands.

The experiment of raising the article of tea in this country has been attempted and abandoned : but never tried in Florida. We are quite sanguine it can be grown there. The northern half of the State lies in the same parallels of

latitude with the noted tea-districts of China; and
the soil is thought to be much of the same character as that on which the choice teas are grown
in that country. Of course it must be at first,
from the very nature of the case, an experiment,
as the culture of any new product, even if seem-
ingly adapted to the locality, is pretty likely
to be; yet, with Chinamen to dress and tend
the plant, its successful culture may be considered
assured. Nor is this an impossible dream; for,
while penning these lines, the telegraph flashes
the arrival in the harbor of San Francisco of one
of the Pacific mail-steamships with twelve hundred Chinamen as passengers. And this whole-
sale immigration is being constantly repeated.
They are now numerous on the Pacific coast; and
we venture to predict that a million of Chinamen will reach our shores within the next decade.
They are industrious and frugal beyond all knowledge of our people; and they will yet solve the
problem of obedient, honest, and respectful labor
at a cheap rate, both in doors and out. Where
now the Green Isle, with its *ever-green* sons
and daughters, hold high and triumphant carnival
let us hope these patient and faithful people

13

may be found instead.   It is a matter for profound regret, that, on the Pacific coast, these " Children of the Sun " are so universally misused, if not often brutally treated by the resident population.   That one of California's senators, himself the son of a foreigner, whose father, under the laws of the United States, was, after a brief residence, admitted to all the rights of citizenship, should publicly denounce these peaceful and law-abiding people as a common nuisance, when their only crime is that of working faithfully, and for one-half the wages of an ignorant Irishman, is too preposterous for credence ; yet the fact seems indubitable.   Shame on this ignoble " Roman."

CHAPTER XIV.

THE most important of all fruits in Florida is the

ORANGE.

There are four varieties to be found in the State. Three of them — the sour orange, the bitter orange, which is very bitter, and the bitter-sweet orange — are found growing wild, in irregular groves, along the principal rivers and streams, though in greatest numbers near the center of the peninsula, up the Ocklawaha River. There, thousands of the wild trees grow; and, in their bearing season, it is a beautiful sight to wander through these natural groves laden with such beautiful globes of gold, peeping on all sides from the bright green foliage, bending low the branches with their weight, and exhaling an aroma at once delicious and powerful. The fruit clings with a good deal of tenacity for a long time after it has ripened; but during the winter and early spring it mostly falls, though you may see the new blossoms and the young oranges while the ripe fruit still remains. No other large fruit-bearing tree does this, to our knowledge; and it is a rare sight, worth a journey, to be able to stand within a wild orchard of these beautiful trees, and feel the drowsing influence of fruit and flowers.

The " sour " variety is too sour for any sort of good purpose; and the " bitter " is too outrageous for any use save to remind you of the " Dead-Sea fruit" whose taste is never forgotten by even those of not usually very tenacious recollections. The " bitter-sweet " are not put to a very great use at present, and may never be; yet the Indians, and some of the native people, use them in the manufacture of a pleasant beverage, quite like our lemonade, and full as palatable perhaps. Barrels of the juice of this orange could be had for the labor of procuring and pressing them. Whether it might be made into a lemon sirup, and thus become an article of commerce, we do not know; but we see no reason it should not, except that our inventive Yankees, or others, have substituted in late years a sirup from chemical combinations, which finds a ready market; and, since the public seem never over nice in small matters, they unhesitatingly gulp it down in quantities sufficient, with other drugs, to in time poison, if it were only taken pure, the whole population.

A very delicious thing is a roasted "bittersweet." They have to be treated with care, lest the oil in the thick rind should get mixed with the juice. If it is ever the fortune of any reader to camp near this variety of the orange, let us suggest that they try roasting one as an experiment. Cut a circular piece of the rind from the top of the orange, push carefully down a portion of the pulp to admit the introduction of sugar, then replace the cap of rind, and fasten with several wooden plugs, not penetrating beyond the rind; then roll it in paper, leaves, or moss, — any thing to keep it clean, — and bury in the earth under the fire. In due time, when it is thought the inside has thoroughly assimilated, carefully withdraw the fruit, and remove the lid; when, armed with a spoon, you have a nectar such as no caterer ever dreamed.

It is generally thought the orange is not a native of this country, but must have been brought to Florida by the early Spanish colonists. This is the accepted theory, but not a known fact. It is our opinion, that the orange is indigenous to the soil; and our reasons are, that it flourishes and seems as hardy and natural to the soil and climate as the magnolia, the live oak, or the palm and date. The great frost of 1835 killed nearly every tree on the peninsula; and yet there is to-day, as good authority asserts, a million of acres of the best lands of the State covered by the wild orange.* Again, the cultivated fruit, the sweet variety, excels the fruit of any other country, both in size and sweetness. This is the general testimony of competent judges, and we unqualifiedly concur, especially when we remember those grown in what is known as the Indian-river country. These seldom reach any of the Northern

* Dr. D. H. Jacques, in the October number of "The Horticulturist" for 1868, gives this statement.

markets, the home consumption being sufficient to require them all. The orange is a native of Asia and the East Indies, and also, we shall add, of Florida. It was not grown in Europe till about the fifteenth century; and it is scarcely likely the Spaniards introduced it, since they cultivated it very indifferently up to the sixteenth century, and Florida was discovered by them in 1512. The fact that some of the early Spaniards left no account of this fruit argues little, since they gave very little account of any thing they either saw or undertook, beyond general expressions. The first elaborate sketches of Florida were given by the Bartrams, father and son, about a hundred years ago; and these writers mention seeing everywhere in the forests, as the traveler may now who seeks the interior and central section, groves of the wild orange, with the fruit so abundant as to actually cumber the ground. The evidence all tends to the one conclusion, so far as we have examined the subject; and that is, as before stated, that the orange is indigenous to the soil in Florida.

Next to the Florida is the Havana orange, a large and handsome fruit of a bright color, with a rich and juicy pulp, and very free from seeds.

The Maltese orange is also a favorite, and is thought the sweetest of all varieties. Its red pulp has been considered objectionable; but now this is less thought of, and already by many it is preferred.* Then there is the Mandarin, a fine new orange, grown in China, which is peculiar in consequence of the looseness of the rind which scarce adheres to the pulp. This is likely to become a favorite.

The orange is the longest-lived fruit-tree known to us. It is reputed to have attained the age of three hundred years, and known to flourish and bear fruit for more than a hundred years. No fruit-tree will sustain itself and produce fruit so well under neglect and rough treatment. It comes into bearing about the third year from the budding, and by the fifth year produces an abundant crop, though the yield is gradually increased by age and favorable circumstances. The early growth of the orange is rapid; and by its tenth year it has grown more than it will in the next fifty, so far as breadth and height are concerned: but it is age that multiplies its fruit-stems; and we are informed, on fair

* The red color of the pulp is said to have come from grafting the orange on the pomegranate.

authority, that ten thousand oranges have been produced by a single tree. A correspondent, in " The Horticulturist" of last year, states that eight thousand were gathered in one season from a tree in St. Augustine.* We know these figures will be thought high, and they are far beyond the common experience; yet they show what may be done. The yield of the present groves in the State, which we observed, would not, however, average over two thousand to a tree; and they were from ten to fifteen years old. These groves have not, as yet, attained maturity; and, in time, the average will run above this figure. They can be marketed readily at twenty-five dollars per thousand: this would give an income of fifty dollars from each tree, or to an acre of rising a hundred trees, five thousand dollars. The labor of one man is quite sufficient to tend the largest grove in Florida, except at the time of gathering, when two are required.

The usual method of propagating the orange is quite like that of the peach; and one who understands the latter, need have no fear of mastering the other. Indeed, it is an easier thing to grow

* See October number of " Horticulturist" for 1868.

the orange in Florida than the peach in New Jersey ; and that is saying considerable in favor of the orange.   It is almost sure to live in the transplanting, and is in every way a very easy tree to manage.   It has but one enemy in the insect line ; and latterly that has not done the slightest damage, and never has the power to destroy the tree, as the " borer " has the peach.   Then the tree itself is free from disease ; and we all know the peach is subject to the " yellows," which, during the last fifteen years, has nearly driven it from every homestead in New England.   Still another great consideration in favor of the orange over most fruit-trees is the tenacity with which the fruit clings to the tree after it is fully ripe.   Ten days is about the usual time for the peach-tree to hold its fruit after commencing to ripen : though varieties differ, we might say the limit was not over twenty days; and all know how a thunder-gust will, when they are ripe, despoil the tree of its fruit.   They have to be marketed rapidly, or they are lost from decay; whereas the orange will, when fully ripe, remain fresh and sound upon the tree for several months, while the winds and storms but gradually shake them off; and when they are thus shaken or gathered regularly, they may be kept for many weeks, though, of course, they slowly deteriorate in size, and lose something in juiciness and fine flavor.

The orange is not only a prolific, but is likewise a steady bearer. It puts forth blossoms during the last half of winter and the first half of spring, by which time the flowering is mainly over: this is itself unusual, and peculiar to this tree. The peach, cherry, plum, quince, pear, and apple bloom promptly and fully, not holding in flower over ten days, and their fruit progressing in growth uniformly; while the orange may hold in February, blossoms, the green fruit, and the matured. These advantages, in connection with its longevity, make it the most profitable and least expensive fruit grown in America.

The ground needs only to be cleared of timber and under-brush to start an orchard of these trees; though, of course, if thoroughly prepared and cultivated, it is only so much the better. The wild-orange tree is used for budding or grafting in the establishing of an orchard or grove. They can be had in any number; and there are parties who are in the habit of procuring them for customers, while others, owning extensive tracts where natural groves abound, supply the market in very much the same manner as the nursery-men North do, and for the present at very reasonable prices, — the usual charge being fifty cents per tree. They vary in size from one to four inches; though a greater percentage of the larger ones die, and are not consequently so desirable. If, however, the tree is in the spring taken up with ordinary care and seasonably transplanted, ninety per cent will live, and become well established the first season. They should be set in rows, about twenty feet apart each way; though the most of the groves we saw were standing nearer. This would give something over a hundred trees to the acre, and in time the whole surface would be covered by the branches. Meanwhile the space between can be used for the growing of vegetables, with actual benefit to the trees.

All trees for transplanting should be cut off to within five feet of the ground, all the main limbs being lopped off, leaving little else beside the stump, and perhaps one or two of the smaller branches.

Those who have groves of the wild orange, in

locations to suit, do not disturb such as they wish
to remain, but take up all others, and with them
fill in here and there to perfect the rows ; and with
the surplus either extend their orchard, or dispose to such as may desire. The usual plan is to
bud the sweet orange on the stock of the wild[1]
orange, at any time after the tree, having been
transplanted, puts forth strong shoots, showing a
renewal of the flow of sap. This is usually done
during the summer ; but budding is attended with
success whenever the sap is flowing upward, and
this is the case most of the year. The bud is
placed in the arms, or trunk, or the shoot when
old enough. If grafting is preferred, you have
only to follow the usual custom; that is, when
strong shoots have started, cut off the tops of the
arms or trunk, and insert the grafts, leaving a
shoot, if convenient, to assist the flow of sap upward, covering the surface cut with any suitable
material, that the air and rain be kept out. The
best plan, if grafting is not resorted to, is, we
think, to bud the shoots after they have had ten
or twelve weeks start, breaking off, early, all shoots
not designed for budding.

The practice of raising trees from the seed is

receiving attention; and, where parties can wait, this is the more desirable plan, still, a much longer time is required to come to bearing. The seed of the wild orange may be planted, though the sweet-orange seed is the best. Both require budding, which, of course, is the only proper way of treating young stock; yet it will be found that the latter will produce a superior fruit.

The sweet variety may be grafted or budded upon the stock of the lemon not only without deterioration, but it is claimed by some to be improved both in size and quality. It is used whenever it can be obtained, the same as the wild orange. It is asserted that the lemon is a much shorter lived tree than the orange; but of this we could not say.

The common impression of those who have never been to Florida is, that this beautiful fruit-tree may be everywhere seen and that they literally abound. This notion would vanish were they to visit the State. It is only in the older settlements that you see groves of the sweet orange. It is true, single trees, or a half-dozen trees, are not unfrequent in numerous localities; but we have never seen over a half-dozen that were worthy of the name of grove or orchard. One at Mandarin, three in St. Augustine, two in and near Palatka, and one, the finest in Eastern Florida, about thirty miles south of New Smyrna, near Indian River.* Western Florida has, in the aggregate, a great many orange-trees; but we are not familiar with any groves there, excepting one at Fort Myers and another at Sarasota Bay.

The tree itself is of handsome form, seldom over twenty feet in height, and near twelve feet across the branches or top, which is conical in form and not unlike the well-trained dwarf pear-tree in general outline, though growing much larger than they do.

The dryer hummock-lands suit the orange best; yet it does well in every soil in the State with which we are acquainted, excepting the scrub-oak land, and even on that it will produce fruit.

The lemon, lime, and citron, all members of the same family with the orange, attain perfection, and are quite as easily grown as the orange itself. They will be found growing wild in Central and Eastern Florida, and may be cultivated with

* Likewise one at Enterprise.

profit. Indeed, whatever has been said of the orange, as to soil and its capabilities, is nearly equally applicable to these; and, beyond this simple mention, we shall therefore not dwell upon them.

Wild grapes of several grades are found in the swamps, and along the rivers and streams, through all the State. They are indigenous to the soil: and it is quite probable that the Spanish people brought with them their cultivated varieties; and these growing wild, and mingling with the natives, have produced kinds varying from every other, — some being quite desirable, and all producing a fair wine-making fruit. Bartram speaks in their praise, in his time; and he was undoubtedly a connoisseur in all matters of that sort. The

SCUPPERNONG GRAPE,

quite famous for its fine wine, may be raised in great perfection and profusion in almost any portion of East Florida. Even the pine lands of poorest quality suit it; and more wine can be made from an acre of this land than from any two acres of the ordinary wine-growing country in Europe. The highest yield in Europe is not over five hundred gallons to the acre : whereas, in this State, over fifteen hundred gallons is a common yield. When these facts become fully known, it must excite the grape-growing population of worn-out France and Italy to try their fortunes in our more favored land.

The Scuppernong is one of the hardiest varieties grown ; is free from the rot, the premature falling of fruit, and is here, quite out of the range of frost. The heat being uniform, and not so great even as it is subjected to in higher latitudes, the fruit matures fully; being a never-failing bearer, and grandly luxuriant in its growth and development. A vine once planted, and well-established, will take care of itself, requiring no pruning or attention beyond giving it an arbor large enough to spread itself upon : with this it is content. This grape is not grown north of Virginia with success, though it may be cultivated in that State ; but it increases in size and in fineness of flavor as you go southward. A gentleman in South Carolina, near the center of the State, has several acres of this grape, and his wine finds a ready market. The banks of the St. John's, we predict, will, in time, be as famous

14

for its vineyards and wine as are those of
the Rhine in Europe. There is no sort of
necessity for America importing the miserable
adulterations that she is now doing, and sending
her gold across the ocean by the millions in
exchange, when, within her own boundaries,
she is favored beyond all lands in soil and climate
adapted to the production of superior wine, and
in such quantities as not only to supply the home
demand, but have a surplus for exportation.

PEACHES

are another crop that can be raised with great success. They ripen in June, and the different varieties carry the bearing season into October. At
present, not much attention is given them; and
that is not so extraordinary either, since few things
have received much attention in Florida for the
past hundred years: and, all the older marks of
civilization having ceased to be, the elements have
almost as clear a title now as on the morning of
creation.

Peaches will bring as high a price in the markets as oranges: and though not so long-lived or
as hardy a tree, yet it will bring quicker returns

to the settler; and their cultivation is so easy a matter that they can not, in due time, fail to command the attention of the shrewd and enterprising. Perhaps the great trouble in this State is, that, as remarked before, it is "so easy a matter" to do this or that in husbandry, that most things go undone, and little stimulus is given labor where its necessity is not absolute.

Figs grow freely, as also the pomegranate and the guava. Our Northern markets are now wholly supplied by importations. The latter article, prepared in a jelly, finds a ready sale everywhere, and is delicious.

Cuba has heretofore furnished us this fruit, but Florida is even better adapted to it than any of the West-India Islands. It is only that it is a new-sounding and strange fruit which will likely prevent its extensive introduction at present. By and by, we may hope for a wider and a more profitable horticulture, not only on the peninsula, but in all of the more Southern States.

CHAPTER XV.

WERE the lands of Florida ten feet higher than
they now are, its future would be great. No State
could then equal it in capacity to maintain a dense
population. As it is, ditching and draining will
have to be extensively resorted to. Three crops
can be taken from the same land in a single year
by a judicious system of rotation.

Many inquiries are made as to the price of lands.
The State owns a large portion of the wild lands,
and the price, we believe, is fifty cents per acre.
At an auction of several thousand acres of unimproved lands, held at Jacksonville in April of this

year, the prices ranged from twenty-five cents to fifty dollars. Lands fronting on the St. John's River between Jacksonville and Palatka, unimproved, can be had from five to fifteen dollars, while improved lands — farms — are held at from twenty to thirty dollars. Of these there are, however, but very few, — scarce a dozen, all told, from one end of the river to the other. Again, wherever there is an orange-grove, the price is enhanced very greatly ; but of these there will not be found a half-dozen, in bearing, along the whole river.

There is an association known as the Florida Land Company, with an office in New-York, who own several hundred thousand acres lying along the line of the Florida Railroad. They are perhaps better prepared to give titles, and offer a wider choice in location and quality of lands, than any other parties. To actual settlers they offer very great inducements, charging little, if any thing, in order that the local trade and travel may be increased along the line of their railway, seeking by this method to obtain remuneration. Their policy is dictated in wisdom, and at the same time is of the utmost advantage to the settler, and especially to those with limited means. Their lands are adapted to the growth of the long-staple cotton, so valuable in all markets.   Sugar-cane and coffee likewise flourish, as also the orange and olive on these lands.

Their line of railway extends from Fernandina on the Atlantic in a south-westerly direction to Cedar Keyes on the Gulf coast, affording unusual facilities for shipping or receiving any article of commerce, since lines of steamers already make prompt and frequent connection at either end of the road with all parts of the country.   It is an important road to Florida, but scarcely less so to the Union, since, by any means should we lose the " Key " to the Gulf, this line of railway would afford a quick and sure means of communication to all of the Gulf States.

The true way for immigrants seeking a home in Florida is to combine, and send out an agent to investigate and report.   Any purchase should be made with a view to the establishment of a colony. In this way they will be able to render to each other great assistance, and can at the same time make their investment in lands to better advantage then if each was to buy or locate on a small tract independently.   They, as a body, could do many things impossible for the individual.

The cost of clearing lands varies from five to thirty dollars per acre, depending upon the kind of lands and manner of clearing. The cheapest is that of clearing the pine lands; and the manner of doing it consists in simply girdling the trees and cutting away the under-brush: the following year there remains nothing but the trunks and dry limbs, which offer no impediment to the rays of the sun sufficient to hinder the growth of any crop. The more expensive clearing is the hummock-lands, where the whole has to be cut down and burnt, and many of the smaller roots removed.

Every company of settlers, who can, should provide themselves with material for a tent. This would give them shelter while preparing their new homes, and be both desirable and inexpensive. Living in warm climates is, of course, attended with some disadvantages; but the dread of reptiles and poisonous insects is not well-founded as to Florida, however true it may be of countries in lower latitudes. Excepting a single harmless water-snake, we saw no reptiles.

Musquitoes, in the season for them, and in their favorite haunts, abound, as they do in many local-

ities at the North. The greatest pest, however,
is the flea. Some they do not annoy, others
escape them entirely; but this was not our fortune. Their abundance is partially explained
when we remember Josh Billings's assertion, that
a careful estimate gives a total of but sixteen
deaths per annum, and these from accident! We
unhesitatingly indorse this statement. Think of
thirty millions of people, with industrious habits,
being able to overtake only that number! Beyond stating that the flea crop of Florida was a
full average one the season of our travels, we
shall dismiss the matter. It isn't a pleasant one
to dwell on. Very pious people should travel in
some other country.

We have, in a previous chapter, adverted to the
settlement of several Northern colonies along the
St. John's, and particularly to that at Palatka, the
pioneer and leading spirits of which are, we .be-
lieve, Mr. H. O. Woodruff and the Rev. P. P.
Bishop, both formerly active and influential gentlemen of Auburn, N.Y. By way of indicating
the interest in, and travel to, the State of Florida,
as manifested within the last few years, we may
mention the fact, that over thirty ladies and gentlemen, from the place above mentioned, sat down

at dinner together in Palatka, one day last spring.

Negro labor seems abundant, yet we understand there is difficulty in securing it. But there are reasons why this is so. They dislike, as a rule, to do any labor for their old masters, since that would seem to them very much like the old system which they now have such a horror of. Again, we suspect the chief reason why the negro is loth to labor is the uncertainty of his wages. The Southern country, immediately after the war, was utterly destitute in every thing, especially of money, and finding they received nothing but promises, the negroes naturally became idle. A couple of crops of cotton has put in circulation among these States considerable currency; but this was not true of Florida, since her crops have been small, and little of any thing for exportation. Hence the colored people have been compelled to give their time to building homes for themselves wherever they were permitted to purchase lands and live in quiet. We say "permitted;" for that more nearly expresses the case than any other term. It seems almost a resolution among the whites not to dispose of any land to the colored people, however much they may be anxious to sell. They are put off with promises; and, in case of sale, they are charged two prices. It is the one thing dear to the late slave population, a home. They seem very anxious to secure to their families a resting-place, where they shall be free from molestation. We met an aged negro, living on Black Creek, a basket-maker, who was working in the shade of a bay-tree, close beside a rude cabin, the temporary home of himself and family. Approaching, we entered into conversation, and learned, from quivering lips, his history and present circumstances. His story was a touching one. He had toiled for his master half a century, in the broiling sun and chilling winds; the war had left him a free man : but he was now aged and infirm, and the fruit of his long life of toil was beyond his reach. He had rented this log-cabin, without floor or chimney, and gathered into it his family, and was struggling to secure a home of his own. He had commenced life anew, and at a time when most of us end ours. He longed to purchase one of God's acres, where he could build his castle, and read his title clear. He did not mind paying

Shylock his price, if the acre could be had. Tears filled the old man's eyes, and his arms grew nerveless, as he repeated his doubts and fears. His family had gathered round, and the wife, an intelligent, thoughtful woman, with eyes fastened intently on the ground, trying, as it were, to solve the mysteries of Providence, in human affairs. Looking out and around us, upon the worn-out and unimproved lands, stretching miles on either side, with scarce a human hand to tend them, yet here was this worthy, industrious, hard-working, and native citizen, half denied the right, in his new condition, to a home among those who recently periled life that he should not go away!

This is but one of many cases coming to our knowledge.

There are no people on the globe so kind hearted, and so cheerful under wrongs and affliction, abiding in the faith of an inscrutable Providence, which, if not understood, is believed to work all things for good, and no people so deserving of considerate and generous treatment from those who were their late masters, as are the negroes of the South.

The social condition of Florida is much like

that of other Southern States. The war has broken up not only families, but neighborhoods, and even States; and what is left is but a wreck of former and more prosperous days. The great questions of State rights and slavery have been settled adversely to the South, and there is no appeal from the decision rendered. Of this fact they seem duly conscious; and having thrown every thing into the contest, and lost, they now find themselves destitute, with no time and less inclination to enter into discussions of what are nothing but dead issues.

At the close of the war, they were not long in taking account of stock, or of coming to the true solution of their situation. Minus friends, capital, and labor, they wisely concluded to accept the situation, and we believe have done so in good faith, and resolved to put behind them their false leaders, together with their principles, which so led them into the red sea of difficulty. In our travels from New York to and through Florida and home again, we met with but a single Southerner who offensively obtruded his politics into our notice, and who still held that the North was the aggressor. This party we met in South

Carolina, and he was of the same "chivalrous" type as might have been encountered at any cross-road in South Carolina before the war. What ditch he could have been hid in when Sherman passed that way is beyond comprehension.

The spirit of the people, taken as a whole, is most excellent, and much better than they have the credit for. There need be, we believe, no apprehension of molestation on this score. All seem intent on getting a living. It would, how-ever, be nothing strange, even with a people disposed to let by-gones be by-gones, if here and there were found uncontrollable characters.

Much time must necessarily elapse, under favorable circumstances, before all classes, so recently antagonistic, and whose relations have been so abruptly and greatly changed, can quietly assimilate, and, out of the chaos of State and National legislation, with the newly-acquired rights and circumscribed privileges of the two races, settle down together in perfect peace and harmony. It would seem marvelous to see a privileged class, a compact people with absolute authority, shorn in a day, as it were, of all power and reduced politically to a common level with their late slaves, and not expect a ripple of rebellion. And yet this marvelous sight is seen in the South. The late elections in the Southern States, and especially in Virginia, attest these facts.

So far from there being any animosity toward the immigrant, the people of the South, on the other hand, seem to welcome every new-comer; and in Florida there is a bureau established to supervise and induce an influx of people from any and every quarter. This is their only resource to build up the State; and in time, with wise legislation, this can be done. A period of prosperity must then dawn on that portion of our country, such as hitherto they have never seen and little dreamed of.

GREAT SOUTHERN
FREIGHT & PASSENGER ROUTE
VIA
SAVANNAH, GA.,
FOR
FLORIDA,
AND ALL POINTS IN THE SOUTH AND SOUTH-WEST.

One of the following First-Class Steamships will sail every other day as follows, — punctually at 3 o'clock, P.M.: —

EVERY TUESDAY,
From Pier 16, E.R., foot of Wall Street,

LEO & VIRGO, of Murray's Line.

MURRAY, FERRIS, & Co., Agents, 61 and 62 South Street.

EVERY THURSDAY,
From Pier 36, N.R.,

HERMAN LIVINGSTON & GEN. BARNES, of Atlantic Coast M.S.S. Line.

LIVINGSTON, FOX, & Co., Agents, 88 Liberty Street.

EVERY SATURDAY,
From Pier 8, N.R.,

SAN SALVADOR & SAN JACINTO, of Empire Line.

W. R. GARRISON, Agent, 5 Bowling Green.

Making close connections at Savannah with the Central R.R. of Georgia trains twice a day for all points in the South and South-west, and with the Atlantic and Gulf R.R. trains twice a day for all points in Florida; also with steamers Nic King and Lizzie Baker, for points on St. John's River.

On and after this date, the Rates of Passage between
New York and the following places will be as follows:—

Savannah...............................$20

Augusta................$22.00	Eufaula................$35.00
Macon.................. 27.00	Montgomery........ ... 37.00
Atlanta.................. 27.50	Mobile.................. 45.00
Columbus.............. 35.00	New Orleans............ 48.00
Albany.................. 35.00	

AND VIA ATLANTIC AND GULF R.R. TO

Jacksonville............$25.75	Enterprise $33.75
Hibernia.................. 28.75	Lake City............... 30.25
Green-Cove Springs.... 28.75	Quincy.................. 35.75
Picolata.................. 28.75	Monticello 33.00
Orange Mills............ 28.75	Tallahassee............. 34.25
Palatka.................. 28.75	

Including meals and first-class state-rooms on steamships.

Passengers for St. Augustine purchase tickets to Picolata, thence by
stage four hours.

On comparison, this will be found to be the cheapest as well as the
most delightful route to the points above named; the above prices to
many points being **forty** or **fifty per cent** less than the All-Rail Route;
the time made being nearly as short, with the addition of a table furnished with all the luxuries of the season, without additional expense.

Trains leave Savannah morning and evening. Sleeping-cars on all
night trains.

Through Rates of Freight given, and through Bills of Lading signed to
all points.

For further particulars, freight or passage, apply to

MURRAY, FERRIS, & CO.,
61 and 62 South Street.

LIVINGSTON, FOX, & CO.,
88 Liberty Street.

W. R. GARRISON,
5 Bowling Green.

NEW YORK, Sept. 15, 1869.

CATALOGUE OF BOOKS AND MAGAZINES

PUBLISHED AND FOR SALE BY

WOOD & HOLBROOK, 15 LAIGHT ST., NEW YORK.

----◆----

HERALD OF HEALTH, and Journal of Physical Culture.

Though professedly a health journal, it takes the broadest meaning of the word health, and includes in it all that relates to physical, moral, and intellectual improvement. Its mission is to teach mankind how to live, so that ignorance, misery, suffering shall be done away, and the reign of health, beauty, and perfection of life shall come in. We want to see the day when health shall take the place of sickness, strength the place of weakness, beauty the place of deformity, and good habits the place of evil ones; temperance substituted for intemperance, a good life for a bad one, and ignorance of one's self be unknown. We ask all who read this to help us, by sending in their own subscriptions and as large clubs as they can, at an early day. Dr. Dio Lewis says, —

"DEAR DR. HOLBROOK, — ' The Herald of Health' delights us. The friends of Physiological Christianity must make it in circulation what you have made it in scope and spirit—cosmopolitan. I am deeply gratified with its broad, genial, catholic spirit. It is pleasant to find ' The Herald' lying side by side with the first-class magazines on our New-England centre-tables.

"With warm esteem,   DIO LEWIS."

From the Scientific American.

"' The Herald of Health' is a journal which contains more sensible articles on subjects of a practical moral bearing, than are to be found in any other monthly that comes to our sanctum."

From the New-York Tribune.

"' The Herald of Health' well sustains the high standard which it has held forth since the commencement of the new series. In fulfilling its task as a ' preacher of righteousness' in the department of Physical Culture, it enjoys the aid of numerous sound thinkers and able writers, whose contributions give popular interest and permanent value to its contents."

$2 a Year; 20 cts. a No.; 4 Subscribers, $7; 10 do., $15.

For $3.35, we send "The Herald of Health" and "The New-York Weekly Tribune" one year. For $3, we send "The Herald of Health" and "The Agriculturist" one year. For thirty subscribers, and $60, we give a Grover & Baker Sewing-Machine, worth $55. We also give Appleton's American Encyclopædia, in twenty volumes, worth $100, for sixty subscribers, and $120. For $5, "The Herald of Health" and "The Atlantic Monthly," or "Harper's Magazine;" price of each, $4.

Physical Perfection.

Containing Chapter on the Structure of the Human Body; the Perfect Man and Woman; the Temperaments; Laws of Human Configuration; Embryology; Childhood; Effects of Mental Culture; Moral and Emotional Influences; Social Conditions and Occupations; Effects of Climate and Locality; Direct Physical Culture; Practical Hygiene; Womanhood; the Secret of Longevity; the Arts of Beauty; External Indications of Figure, &c., &c. Beautifully illustrated with one hundred engravings, and handsomely bound. Price, by mail, $1.50.

The New Hygienic Cook-Book.

BY MRS. M. M. JONES.

This work contains several hundred recipes for cooking the most palatable and wholesome food without the use of deleterious compounds. Also, directions for Washing, Ironing, removing Stains, Canning Fruit, &c. Price 30 cents.

The Turkish Bath; its History and Philosophy.

BY ERASMUS WILSON.

With Notes and an Appendix by M. L. HOLBROOK, M.D. This is the only book on the Turkish Bath published in this country. Illustrated. Price 25 cents.

Woman's Dress.

With numerous engravings, showing how woman's clothing can be made beautiful, healthful, and comfortable. Price 30 cents.

The Tree of Life ;

Or, Human Degeneracy, its Nature and Remedy.

BY ISAAC JENNINGS, M.D.

It contains the following Table of Contents : — Man's Spiritual Degeneracy; Physical Depravity; Physiological and Psychological Reform; Remedy for Man's Spiritual Degeneracy; Man's Physical Degeneracy; Constitution of Human Physical Life; Vital Economy, or Organic Laws of Life; Source and Mode of Transmission of the Principle of Human Physical Life; Predisposition to Disease; Hereditary Diseases; Mode of Renovation of Impaired and Feeble Vital Machinery; Analysis of a Few of the Most Prominent Symptoms of Disease; Law of Contagion and General Causation; Medical Delusion; Remedy for Man's Physical Degeneracy ; General Directions Resumed ; Specific Directions, — Croup, — Dysentery, — Cholera; Final and Effectual Remedy for Man's Physical Degeneracy. Price, by mail, $1.50.

Physical, Intellectual, and Moral Culture;

OR,

HOW TO LIVE WELL.

By F. G. WELCH,

PROFESSOR OF PHYSICAL CULTURE IN YALE COLLEGE.

This is the most extended and perfect work of its kind ever printed. About one-third of the book is devoted to an explanation of Gymnastics, as taught by Prof. Welch in Yale and other colleges. Another third is devoted to health-culture, and the improvement of the intellect, and such matters as all should know; and one-third to moral culture.

Every man and woman who would make the most of life should have a copy.

Price $2.25, by mail.

A Winter in Florida;

OR,

OBSERVATIONS ON THE CLIMATE, SOIL, AND FRUITS OF THE STATE; WITH HINTS TO THE TOURIST, INVALID, AND SPORTSMAN.

By LEDYARD BILL.

This is the most complete work on Eastern Florida and the famed St. John's River yet published. It contains sketches of all the towns and cities along the St. John's, with a detailed description of the scenery, and the wild game which inhabits the country. All contemplating a tour to Florida should possess it.

About 225 pages, 12mo, with four illustrations. Price $1.25, by mail.

Any or all of the above works will be sent free, by mail, on receipt of the price. Address,

WOOD & HOLBROOK,

13 & 15 Laight Street, New York.

THE NEW-YORK HYGIENIC INSTITUTE,

Nos. 13 & 15 Laight Street, New-York City.

A. L. WOOD, M.D., Physician.

The objects of this institution, which has been in successful operation for more than twenty years, are two-fold, viz. : —

1. The treatment and cure of the sick without poisoning them, by Hygienic agencies.

2. To furnish a pleasant, genial HOME to friends of Hygiene throughout the world, whenever they visit this city.

CURE DEPARTMENT.

Thousands of invalids have been successfully treated at this institution during the past twenty years, and its fame extends wherever the English language is spoken. Its appliances for the treatment of disease without the use of poisonous drugs are the most extensive and complete of any institute in America. They comprise the celebrated

Turkish Baths, Electric Baths, Vapor Baths, Swedish Movement Cure, Machine Vibrations,

the varied and extensive resources of the WATER CURE, LIFTING CURE, MAGNETISM, Healthful Food, a Pleasant Home, &c. Particular attention is given to the treatment of all forms of CHRONIC DISEASE, especially of Rheumatism, Gout, Dyspepsia, Constipation, Torpidity of the Liver, Weak Lungs and Incipient Consumption, Paralysis, Poor Circulation, General Debility, Curvature of the Spine, Scrofula, Diseases of the Skin, Uterine Weaknesses and Displacements, Spermatorrhœa, &c. Any one wishing further information, should SEND FOR A CIRCULAR, containing further particulars, terms, &c., which will be sent free by return mail.

BOARDING DEPARTMENT.

We are open at all hours of the day and night for the reception of boarders and patients. Our location is convenient of access *from* the Railroad depots and Steamboat landings, and *to* the business part of the city. Street cars pass near the doors to all parts of the city, making it a very convenient stopping-place for persons visiting the city upon business or pleasure. Our table is supplied with the BEST KINDS OF FOOD, HEALTHFULLY PREPARED, AND PLENTY OF IT. In these respects it is unequalled. Come and see, and learn how to live healthfully at home. Terms reasonable.

Drs. WOOD & HOLBROOK, Proprietors.